2003

Transparent Oceans:

The Defeat of the Soviet Submarine Force

by

Louis P. Solomon

ISBN 0-9728085-1-5

Dedicated to the members of the LRAPP community

Preface

In 1967 the Office of Naval Research established the Long Range Acoustic Propagation Project (LRAPP). The purpose of LRAPP was to solve the problem of antisubmarine warfare (ASW). At that time, the Cold War between the United States and the Soviet Union was in full swing. If the Cold War dissolved into actual combat, there was a distinct possibility that the United States of America would be destroyed. The defeat of the Soviet Union was the paramount objective of the United States government. Many organizations contributed to the defeat of the Soviet Union; the LRAPP organization made a measurable contribution to the final victory.

LRAPP consisted of Americans from government, military, and industry. They worked together over a 25 year period to defeat a military and political force that threatened the very existence of the United States. This team of superbly capable individuals operated with vision and resolution, remaining focused on the problems associated with Antisubmarine Warfare. This team was supported by senior people in the U.S. Navy, both military and civilian, who understood the ASW problem and realized that this team was perhaps the only real potential solution to the antisubmarine warfare problem. The LRAPP organization overcame technical, managerial, and bureaucratic problems that in the beginning were considered insurmountable and yet were vanquished by them during this two-decade span.

The general public—who support the Navy's people and programs—should feel a sense of satisfaction that their taxes were spent intelligently on their behalf. They should feel pride that several dozen American citizens with unusual talents worked together to accomplish a goal that directly contributed to the protection of the general public. Government leadership is not always perfect, but it did understand the nature of the threat to the United States and Western Civilization. In the case of LRAPP, the government had the courage, brains, and leadership to generate a team, support it, and allow it to be led intelligently and with purpose.

Acknowledgments

I received substantial help in writing this book. The members of the editorial board who contributed their time and constructive criticism are Roy D. Gaul, Kenneth W. Lackie, Bob Spindel, Darrell Milburn, Dick Van Orden, Wentworth Williams, Ken Osborne, Terry McCloskey, Wendell Sykes, Ed Chaika, Paul Bucca, Nancy Bedford, Glynn Ellis, Charlie Stuart, and Claude Brancart. Their assistance was invaluable. I also thank the entire LRAPP community for their participation, enthusiasm, and kind efforts to assist me while I was writing this book. I particularly want to thank the LRAPP members who took the time to write the vignettes in the text. These vignettes will bring back memories of the program and the people who made it successful.

In a book that attempts to tell historical events, three issues demand careful attention: the review of the factual material; the organization of the book and how the story will be told; and the style in which it is written.

- For authenticating the factual material, I depended substantially on the editorial reviews of the men and women acknowledged above.
- The organization and integration of the myriad details, which may conflict because memory is sometimes misleading, was undertaken and performed by the editor, Linda Jenkins. She integrated the information and helped substantially with the flow of the material. It is to her that a great amount of the credit belongs.
- The style in which the story is written is entirely my own. While many people assisted me with their comments, both oral and written, the responsibility for this entire work is mine and mine alone. I bear the full responsibility for any errors that might occur.

Of special note is Appendix C. Appendix C includes photo-graphs that were contributed by Paul Bucca, Darrell Milburn, and Nancy Bedford.

Finally, I thank the United States Navy for granting me the privilege to participate in many different roles throughout the life of LRAPP. The real credit belongs to the Navy in supporting LRAPP to the fullest extent possible, allowing LRAPP to integrate science and technology with naval operations. The integration of those capabilities provided the solution to the antisubmarine warfare problem. The ultimate recipients of the Navy's vision and determination were the citizens of the United States of America and, in a larger sense, the entire Free World.

Contents

CHAPTER ONE
Introduction

The Threat[1]

The advent of the nuclear submarine was a formidable new threat to the Free World. It was much faster than earlier submarines, much quieter, seemingly undetectable, and carried substantial and continually improving weapons. Essentially two types of submarines were developed: the attack submarine and the missile-carrying submarine. The attack submarine was designed to attack merchant ships, surface warships, and enemy submarines. It had a long history of success in both World War I and World War II. The supply life line between the United States and England was almost severed by German diesel submarines during both of the World Wars. The Japanese merchant traffic within the Greater East Asian Co-Prosperity Sphere was almost completely destroyed by American diesel submarines operating in the Western Pacific during WWII. The development of the nuclear submarine made the attack submarine more deadly than its predecessors. The Soviet Union, with the largest submarine force in the world, clearly had in its hands a potential major threat to free world commerce.

The missile-carrying submarine was designed to position itself in the ocean and, upon receipt of the appropriate command, launch its missiles quickly against a previously designated target. Initially, the missiles were not armed with nuclear warheads and could only travel short distances (approximately 100 miles). In a short period of time, the application of advances in technology allowed the missiles to travel much longer ranges and carried nuclear warheads. The mission of nuclear ballistic missiles is to destroy large areas, major installations, or cities.

All ballistic missiles are powered by their engines during the initial phase of their flight path. They lift off from their launch point, and fly, under power, into the atmosphere. At some point, their engines shut off, and the missiles follow a path that is influenced primarily by the force of gravity. This path is a ballistic trajectory. The path of a ballistic missile is very similar, except in

---

[1]See Appendix D, Glossary, for a definition of all technical terms

scale, to that of a thrown ball. The path followed by both the ball and the ballistic missile is determined by their initial angle to the horizontal and their initial velocity. These two properties uniquely determine the path. The main external force that is exerted upon the missile (ball) is gravity. This force slowly retards the initial vertical speed and causes the missile to curve over the earth and then return to the ground to hit a specified target. By working the problem backward, we can specify a particular point that we want to hit; then, adjust the required initial launch angle for a given initial velocity.

There is a major problem with the ballistic missiles that are used by the attacking force. Once the incoming missiles are detected by radar or other similar devices, the defense force computers can calculate the trajectory and the speed of the incoming missiles. Anti-ballistic missiles may then be fired to any point along the trajectory, arriving at the desired point at the same time as the incoming missile. Assuming everything works correctly, the incoming missile will then be destroyed. The problem of having two missiles arrive at the same spot in the atmosphere at the same time is not particularly difficult. Developing the radar detection systems, and coordinating with the launch of the anti-ballistic missiles is a difficult, but not impossible technical problem. The entire problem is solvable because ballistic missiles follow predictable paths. While ballistic missiles are extremely fast, radar and computers are much faster.

The development of the cruise missile complicated the problem of missile defense. The path of a cruise missile is not predictable from basic physics. A cruise missile can be launched and then, using internal computer calculations and sensors, fly a path that can be adjusted along the way. The path variations may be pre-programmed or, by using advanced logic and on-board sensors, select between several different possible adjustable paths that lead to several possible end points (targets).

Destroying cruise missiles in flight is extremely difficult. A better and possibly simpler approach was to destroy the missile-carrying submarines before they could launch their missiles. In order to provide a timely defense, it is necessary

> In the early days of the Cold War (circa 1955), Soviet missile submarines had to come close to the United States because the range of the missiles was quite short. With each passing year, the missiles could be launched from progressively farther away. In the last years of the Cold War, Soviet missile boats could launch their missiles from several thousand miles away, thus allowing the Soviets to launch their missiles from waters immediately contiguous to the Soviet Union.

to detect the missile carrying submarines prior to missile launch and be able to destroy them, essentially instantly, on command from the National Command Authority, resident in Washington DC.

The oceans cover approximately 70% of the earth's surface, so US attack submarines had to be distributed throughout the oceans of the world, all the time. The US attack submarines had to detect, localize, classify, and be prepared to kill the Soviet missile carrying submarines on command, in a very short period of time. Aside from the communication issues, which were major in themselves, the problem of anti-submarine warfare became the crux of the missile defense problem.

A Note on Submarines at the Turn of the Century

John Phillip Holland's original invention of the submarine was in 1877. The first submarine purchased by the U.S. Navy was Holland's sixth submarine. It was commissioned as the USS *Holland* and designated the SS-1.[2] Submarines were built by most of the larger seafaring nations. It was recognized quite early that the submarine offered many operational opportunities as a new Navy platform; it was a threat of unparalleled danger. Its development was pursued with great vigor by the world's premier navies.

World War I saw operating submarines become reality. Germany's use of unrestricted submarine warfare was considered "unsportsmanlike." At that time it was still considered that warfare was to be conducted according to time-honored rules. The idea that ships could be attacked without first being given warning was simply considered improper, even if it was highly effective. Until the allies formed convoys, the submarines had the upper hand.

Convoys consist of a relatively large number of ships, sailing together. They can all be protected by a small number of warships. If the convoy is attacked, the warships can provide defense and counter-attack. Also, in the event of an attack, the ships of the convoy scatter in all directions, leaving the submarine only one or two victims. Convoying has always been proven successful, even from the time of antiquity.

Submarines of the World War I era , however, had one particular difficulty: they really didn't work well submerged! True, they could submerge and then resurface, but they really didn't operate well while they were under water. Submarines also had other problems. Their sonars were primitive and

---

[2] http://www.geocities.com/gwmccue/index.htm

the displays even more so. Living conditions on board the so-called "pig-boats" were appalling. Their operating ranges were short, and the favorite method of attack was on the surface. The WWI submarine was really a submersible surface ship that could resurface. Nevertheless, they caused great concern among the allies because of the substantial negative effect they had on seagoing commerce.

The fall of France early in World War II allowed the Germans to acquire and use French Atlantic ports. Use of these ports substantially reduced the transit times to the North Atlantic—consider that St. Nazaire is much closer to the Atlantic than Wilhelmshaven. When the German submarines left Wilhelmshaven they could sortie to the Atlantic from Germany either by traversing the English Channel (highly risky) or traveling through the Baltic, traverse the North Sea, circle Scotland and Ireland, and finally enter the North Atlantic (much less risky, but very long). Departing from St. Nazaire allowed them direct access to the Bay of Biscay, which was almost within striking distance of the Western Approaches to English ports. The use of St. Nazaire allowed much longer times in the combat area.

World War II saw the submarine threat become highly effective, and almost catastrophic for the allies. The equipment developed during the period between World Wars I and II allowed significant improvements in submarine capabilities, range of operations, and living conditions. The principal constraint imposed on World War II submarines was their need to be frequently re-supplied with armaments, fuel, and food. Increasing the size of the submarines allowed them to carry more armaments, more fuel, and more food. Larger WWII submarines could remain at sea for longer periods and range much farther from friendly ports their smaller WWI predecessors.

The speed of World War II German submarines was approximately 12 to 14 knots on the surface, but much slower when submerged. While the quality of the German submarines was much higher than their World War I

> Diesel engines require two items to operate: fuel and air. Thus, it was impossible to operate diesel engines while the submarine was submerged. The submarine was under battery power while submerged. When the snorkel was invented later in WWII, diesel engines could operate while the submarine was submerged by taking needed combustion air through the snorkel. The submarine could then operate its diesel engines while submerged at shallow depths (keel was about 45 feet below the sea surface) with the snorkel raised above the sea

predecessors, they still were essentially surface ships that submerged and resurfaced. They generally attacked their targets on the surface during the night and submerged during the day.

The German submarine operated in one of two modes: on the surface using their diesel engines or submerged using battery power. While they operated on the surface, they replenished their inside air, which became stale very quickly in the confines of the closed submarine, and recharged their batteries.

The United States Antisubmarine Warfare (ASW) forces initially had little capability to detect submarines until an attack took place. The operative phrase was "Flaming Datum." When a surface ship was either torpedoed or fired upon with the submarine deck gun, an attack was clearly underway. This action caused the naval surface escorts to begin an intensive search for the surfaced submarine. Searchlights and binoculars were the principal means of detection. For submerged enemy submarines, the only detection equipment available to the Allied navies was the very rudimentary sonar which had been in use since World War I.

A tremendous step forward was the introduction of surface ship mounted radar, early in World War II. Radar, even its infancy, allowed surfaced submarines to be detected with relative ease. During the WWII era, radar was big and ungainly. It could only be operated from surface ships, but it could detect targets on the water, i.e., surfaced submarines. The rapid growth of the use of radar by surface ships forced the German submarines to begin to operate more time submerged. This requirement to remain submerged for longer periods of time caused their range to be restricted. Their maximum speed was substantially limited, and they had to surface more frequently to recharge their batteries. Operating on the surface became a dangerous thing to do. As a counter to the Allies radar, Germans invented a radar detection system that informed the submarine that a radar was operating in their vicinity. This radar detection system provided some increased warning time to the Germans so they could take evasive action and re-submerge.

Radar continued to develop rapidly. Soon radar was made sufficiently small and light so that it could be used on airplanes. This advancement made life for the submarine even more perilous. While the submarines had a radar detection system, the airplane could detect them sooner and attack much faster than the submarine could escape from the attack by submerging. Even if the submarine could submerge prior to attack by the aircraft, it could only attempt to reach safety at relatively slow speeds (several knots) while the aircraft, which had noted the spot where the submarine submerged, could move at several *hundred* knots.

In the early 1942, things were not going well for the submarine and, in a short time, became worse. Detecting a surfaced submarine, which was a large object, became commonplace and easy. When the Germans invented the snorkel, however, only a relatively small tube broke the sea surface and made detection of a submarine much more difficult.

Allied engineers were set to work on using radar to detect the relatively small snorkel tube that broke the sea surface. They were intent on developing and building a radar system that operated at a particular frequency whose wavelength was about the size of the projecting snorkel. If the reflecting device is about the same size as the incoming wavelength then there is maximum reflection. The development of radar operating at the desired frequency was quickly solved under the pressures of wartime requirements. The importance of solving the detection of snorkeling submarines had life-and-death implications; the German submarine had to be defeated. These pressures, and the contributions of the technical community produced a working radar in a short time, and soon a snorkeling submarine could be easily detected.

This series of events was unfavorable for the German submariners. The detection and subsequent kill rate of German submarines increased alarmingly. Although the Germans had invented a radar detection system for surface ship radars, they never realized that their snorkeling submarines were being detected by airborne radar tuned to detect their snorkels.

The use of radars, both surface ship and air borne, made the life of surfaced German submarine very hazardous. If a German submarine managed to avoid detection prior to attacking the surface ships, once the attack began all parties knew that a submarine was around. After its first attack, a German submarine was forced to submerge. It then found itself under heavy attack from naval surface forces using their active sonars. The sonar transmitter, located under the keel of the surface ship,

> The significance of this discussion is that the submarines were detected in two ways:
>
> - by radar when they were on the surface or sub-merged and using their snorkels, which also broke the surface of the water, and
> - by using active sonar with the returning echo from the acoustic signal reflecting from the submarine.

sent out an acoustic signal ("pinged") and waited for the reflected echo. If no echo was returned, then they sent another ping and the search continued. The ranges between the surface ships and their submarine prey was generally less than 1 mile, or 2000 yards.

The receiving sensors (hydrophones) could be steered mechanically by the sonarman. If an echo was returned from a submerged target,[3] then the sonarman could tell the captain which direction to steer the ship. As the distance between the attacking surface ship and submerged submarine became shorter, the interval between the transmitted ping and the received echo became smaller. At some point, the sonarman informed the captain of the relative position and estimated depth of the submarine, and a barrage of depth charges ("ash cans") would ensue.

This type of detection process has many possibilities of error, such as determining the bearing of the returning echo or the submarine's depth. While it seems quite straightforward to detect and kill a submarine, that was not the case. Substantial skill, much practice and, to some extent, luck, were required.

Many people with different skills were involved in at-tempting to detect, localize, classify, and kill a submarine. Generally, they were all on one surface ship or a single aircraft. In some instances, two or more surface ships worked together to prosecute a submarine. In other instances, a surface ship worked with an aircraft; however, these instances were rare. When two or more surface ships worked together, they were always close to each other. Each ship could see the other, and they had a feeling on how they were acting as a team. They coordinated their activities using radio.

The significance of the ASW problem during WWII caused many organizations to focus on it. At Harvard, a group was dedicated to acoustics. Mr. Ted Hunt, one of the members of that group, made substantial advances in the science of underwater acoustics. Hydrophones were built and tested. Materials were investigated to ascertain their electromechanical properties. Some mathematical work was done on the problem of underwater sound propagation. These focused efforts helped immeasurably in developing the art and science of underwater acoustics and its applications to solving the ASW problem.

After World War II, members of the Harvard group followed several academic disciplines at different universities. Some of the group went to AT&T, where they continued to pursue their efforts in underwater acoustics.

---

[3] See Appendix F, The Basics, for details.

The end of World War II saw the successes of the ASW forces. They had developed and applied scientific methods to several different aspects of the problem with great success. The development of radar had forced the submarines to operate submerged. Short-range, high-frequency active acoustics worked; submerged submarines could be detected at relatively short ranges, e.g., a few thousand yards.

The allied ASW forces were pleased with their success, but the more thoughtful leaders recognized some ominous developments in the submarine world. The Soviets had captured the new Type XXI submarines from the defeated German forces. Those submarines were substantially better than the main submarines used by Germany throughout most of World War II. They could remain submerged longer, they were faster, and they had longer ranges. These leaders realized that a Cold War might develop and that the Soviets would then have a submarine force-in-being. It appeared that the ASW forces could not rest on their laurels; substantial advances in ASW capabilities would have to continue.

Chapter Two
The Cold War and Antisubmarine Warfare

General Comments

The Cold War (1946–1989) has substantial precedent in written history. Tensions between nations have existed for thousands of years, and actual combat between nations has lasted for substantially longer periods than the Cold War between the United States and Russia.

During modern history, however, there have been few protracted periods where the nations of the world have divided themselves into armed camps for such a long time. Further, due to the skill and persistence of the scientific community, weapons of terror were and still are completely capable of eliminating life as we know it on our entire planet.

During the period of extended tension, the United States as the leader of the Free World and the Soviet Union as the leader of the Communist World, faced each other with antagonism and fury that were barely contained. However, as we know, the Soviet camp disintegrated in 1989. The Cold War was declared officially over; the United States and the Free World had won.

But did a specific country win the Cold War? Can a specific organization be singled out as having won the Cold War? A case could be made that the Manhattan Project, which developed the Atomic Bomb, was perhaps the one project that eventually resulted in the success of the allies in defeating Japan during World War II. (Germany had already surrendered by the time the A-bomb was used).

I contend, with a certain amount of tongue in cheek, that winning of the Cold War was materially affected by an organization called the Long Range Acoustic Propagation Project (LRAPP). The LRAPP organization was, and remains, virtually unknown to almost everyone, including almost the entire U.S. Navy.

The differences between the two projects are substantial. The Manhattan Project cost billions of dollars, involved tens of thousands of people operating at secret sites, for a relatively short period of time: 5 years. LRAPP spent considerably less than $300 million, never involved more than

200 people, and existed for approximately 20 years. The Manhattan Project had the support of the President of the United States, as well as essentially unlimited funds and bureaucratic control over its own destiny. LRAPP was unknown to almost everyone, particularly at the senior levels of the U.S. Government, and had no bureaucratic control over its destiny. Yet the comparisons between the two programs, if measured by their contributions to the successful conclusion of World War II and the Cold War, are startlingly similar.

The Cold War was fought on a myriad of fronts. These fronts can be lumped together in such broad generic categorizations as ideological, political, economic, and technical, but not military. Confrontations in the sense of actual military battles did not occur between the principal leaders of the two major camps. Such confrontations were left to their subordinate partners. However, in one military area, and in one military area alone, the United States fought the Soviet Union in a military sense (with the sole proviso that no weapons were actually fired) on a 24-hour-a-day-basis. This battle occurred in the broad area of undersea warfare, specifically in the narrow area of Antisubmarine Warfare (ASW).

From the beginning of the Cold War in the late 1940s until the collapse of the former Soviet Union in 1989, the U.S. Navy fought Soviet Navy submarines on a daily basis. In this titanic struggle, all the resources of both military organizations dueled against each other. At the end of each day, the competing forces knew who had won. Sometimes it was the U.S. Navy, and sometimes it was the Soviet Navy. New equipment was constantly being developed; new tactics were continually being generated and tested under conditions that differed from combat in only one aspect: no shots were actually fired. However, the entire process of detection, classification, localization, and establishment of a fire-control solution was attempted and frequently achieved by each side.

> In the history of the world, many games have been invented. In the military world the concept of games is used for training purposes. Training is given to the individual military person in each of the armed services. Then training is given to larger organizations, increasing in size up to full armies. All military organizations maintain the ancient belief, "You fight as you train." During the severe stress of combat there is no time to become creative and imaginative. In fact, if deviations from established procedures are attempted during actual combat, they invariably lead to disaster.

One result of war games is to establish training procedures that will lead to successful combat, when and if that need ever occurs, against anticipated enemy capabilities. War games themselves, and their logical extensions, military exercises, are held within the military services. In certain scenarios, there are joint exercises where components of different military branches practice together. Generally, however, military exercises are held within a specific service, e.g., a naval exercise is held by designated components of naval operating forces.

In a military exercise, the role of the opponent is played by specially designated forces. These forces use the known or assumed tactics of the opposition; such information comes from the intelligence arms of the military forces. The salient point is that the opposing forces are played by substitutes. It is a constant and major concern among the commanders of the military forces that they are training their forces to fight against opponents who may use quite different tactics if actual combat takes place. They are never quite sure that the military hardware developed and used by the potential enemy will perform or be employed as predicted by the intelligence community. Military surprises are never pleasant, and frequently result in unanticipated casualties to both men and equipment.

None of the military services of the United States during the Cold War ever tested themselves against their Soviet counterparts, except the part of the U.S. Navy that was responsible for undersea warfare. During the entire Cold War, the U.S. Navy participated in operations throughout the oceans of the world against their Soviet counterparts on a daily basis.

The Game

In the history of the world, seldom has there been a game as demanding and exciting as ASW. The game was played by thousands of people, military and civilian, on both sides on a daily basis for the entire duration of the Cold War. The equipment was the finest to be developed by each side. No effort was spared in using national resources to develop a team that could defeat the other. The actual test of the tactics and equipment occurred in daily operations between the U.S. Navy and the Soviet Navy.

As participants in any other complex game, the teams needed training. If actual combat had occurred, there would be no opportunity to try different tactics on another day. The loss of men and equipment would have been catastrophic to the losing side. The fate of the war would be decided and the history of the world would turn to one side or another. The United States national leadership required the U.S. Navy to develop tactics and equipment that could instantly kill Soviet submarines on command.

Early in the Cold War senior Navy leadership recognized that undersea warfare would be one of the keys to the success of the United States in its confrontation with the Soviet Union. Both previous world wars had demonstrated the amazing power of the submarine: the first and most successful stealth platform[4]. A perfect stealth platform is essentially undetectable. A submarine is virtually undetectable, and the quieter it becomes, the less detectable it is. The development of the nuclear submarine with its increased endurance, ability to stay submerged for long periods of time, and substantial speed under water increased its potential impact on the conduct of the war. The development of nuclear missiles that could be launched at a moment's notice from anywhere in the world's oceans from hidden and undetectable platforms was the principal threat faced by the United States.

Essentially a weapon platform, the nuclear missile submarine had evolved so it could hold sufficient weapons that, if used, could obliterate all life on the planet. The missiles were capable of being launched in a very short time—on the order of minutes. The U.S. Navy recognized that they needed to develop a capability that could detect, classify, localize, and prepare to kill Soviet submarines before they could launch their nuclear missiles. Since Soviet submarines could be ordered to launch their missiles at any time from the Soviet Command Center in Moscow, it was imperative that the U.S. Navy have the capability to destroy the Soviet missile-carrying

---

[4] A platform in the Navy is a surface ship, an airplane, or a submarine.

submarines from the instant they left their home port for a 60-day patrol until they returned. Never had such a requirement been levied by any nation on their military forces in the history of the world. The task seemed impossible, so the U.S. Navy considered the problem and immediately adopted a bureaucratic approach.

As a reaction to a deteriorating situation, the Navy formed a new group, the Director of Antisubmarine Warfare (designated as OP-095 within the administrative structure). The first commander of OP-095 was Vice Admiral (VADM) Charles B. Martell, who was personally selected by then-Secretary of the Navy, Mr. Paul Nitze.[5]

A Lesson on How the U.S. Navy is Organized

It is important to understand how the U.S. Navy was organized in 1967. That was the time when the initial events in this story took place. At the present time, to some extent, the Navy continues to be organized the same way. At its foundation, the Navy is organized along *platform* lines. The people who operate such platforms devote their careers to the development, testing, and operation of specific platforms. The platform is what fascinates the devoted officers and enlisted personnel of the Navy. These men (and later women) choose their area of focus early in their career and, with only rare exceptions, remain in their field of specialization throughout their careers. Submarine officers are devoted to submarines. It is unusual for a submarine officer to be associated with aircraft or other platforms instead of submarines. Such exceptions do occur but generally above the individual platform command level.

The key point is that the Navy is organized along platform and not functional lines. All platforms have homes, and organizations that support them. There are schools that teach naval personnel about their platforms. The Navy is focused on understanding the characteristics of its platforms.

But, ASW is a function. It has no home nor supporting organization. The function of ASW is to use platforms of different types. The complexity of the ASW problem is substantial and requires special training to understand the different aspects of the problem, as well as the special needs of each facet of the problem. These different facets can be addressed using the special capabilities of different platforms, but at different times.

---

[5] Interview with RADM Ed Snyder, USN (retired).

VADM Martell intuited specifically that the detection of submarines would have to be based upon the improved understanding of the physics of underwater sound propaga-tion. He turned to the scientific resources within the Navy, which were embodied in the Office of Naval Research (ONR). He recruited Dr. J. Brackett Hersey from the Woods Hole Oceanographic Institution to lead the effort. Dr. Hersey was a recognized expert in the complex field of underwater acoustic propagation.

> The formation of OP-095 was unique within the Navy, and the choice of VADM Martell was fortuitous. He immediately recognized that the bedrock for possible success in dealing with a virtually impossible problem would be science.

VADM Martell needed a project that would be under the direct control of Dr. Hersey, and he convinced Dr. Hersey to come to Washington, D.C., to manage it. The project would organize and develop the scientific tools and information necessary to deal with underwater acoustic propagation over long ranges. Detection of Soviet submarines would initially have to be across both the Pacific and Atlantic Oceans. By any measure, these enormous distances would necessarily entail the understanding of long-range acoustic propagation. The Long Range Acoustic Propagation Project, LRAPP, was established within and reported to ONR; overall tasking and prioritization was received from OP-095.

This tale describes the generation, existence, and demise of LRAPP and its successors. The story tells the complex series of events that took place from 1967 to 1987. It is arranged in roughly chronological order and will describe the role of several professional communities that almost never deal with one another except through coercion or happenstance. To appreciate this complex tale, it will help if the reader has a modest understanding of the bureaucracy of the U.S. Navy and its components, as well as the larger world of the scientific community. In the course of this book, the reader will meet several key individuals who possessed the foresight and determination that enabled LRAPP to have a positive effect out of all proportion to the funds involved. It was the contribution of LRAPP and its people that provided the understanding of the physics that controlled ASW. These people integrated their science and technology with professional sailors, and the two groups, operating together, solved the ASW problem. Finally, the reader will understand the crucial roles of several hundred people, all of whom contributed to the success of LRAPP in helping to win the Cold War.

CHAPTER Three
The Problem

General Comments

World War II was over. The Allies defeated the Axis nations: Germany, Italy, and Japan. The technology of the West had triumphed. However, the world was experiencing an uneasy peace, which slipped almost immediately into the Cold War. Europe was in shambles. The Soviet Union began to challenge the West. The strong land warfare capability of the Soviet Union and their large population made the United States and its Allies in Western Europe restive. Did the Western posture have a problem that made it potentially vulnerable to the Soviets? Yes, at least one: the Soviet submarine threat, which was a fleet in being. In all other military areas, the western Allies, and in particular the Arsenal of Democracy, the United States, had a commanding lead. The era of nuclear parity was still in the future; only the United States had nuclear weapons. But, the Soviet Union had an existing submarine force, which had been enhanced through their acquisition of the existing German submarines, as well as their newest creations that had not played a role in the final days of the war at sea.

Throughout the entire war, the Battle of the Atlantic was fought between the German U-boats and the Allied surface ships. The United States was attempting to send supplies to England and eventually all of Europe. These merchant convoys were attacked and harried by the German U-boats with some substantial success. In his memoirs, Winston Churchill remarks,

> The only thing that ever really frightened me during the war was the U-boat peril.[6]

Churchill was correct. The German U-boats were both active and successful in what they wanted to accomplish. The issue of Britain being able to sustain itself was in doubt through the early years of the war. At one

---

[6] Winston S. Churchill, *Memoirs of the Second World War*, Houghton Mifflin Co., 1959, p. 410.

point in 1943, Great Britain had only about six weeks of food available to feed the entire population.

German U-boat operations from 1939 to 1945 were disturbing. In the Atlantic, a total of 1,133 U-boats were commissioned; during the conflict, 782 were sunk. The submarine service had the highest percentage of losses incurred by any branch of the military service (true on both sides). The loss in shipping was interesting and also quite troubling to the U.S. Navy, who reviewed this information after the war was over. From 1939 to 1943, the German Navy sank substantially more merchant ships than were built by the combined efforts of the British Empire and the United States of America.[7] The favorite method of attack by the so-called German Wolf Packs was a surface attack at night and an attack at periscope depth (approximately 60 feet) during the day.

Submarines of that era were really surface ships that had the unusual characteristic that they could submerge and resurface. Diesel engines, the main source of energy for the submarines, could only be used if there was a supply of air to feed the engines. Diesel engines require both air and fuel to operate. So the diesel engines could only be used when the submarine was surfaced. When the submarine was submerged, their source of power was batteries. Battery technology at that time was not advanced, and the amount of power available was rather limited. Submarines were quite slow while submerged and had limited endurance due to their dependence on batteries.

The on-board living conditions were trying. Two-minute showers were allowed on board U.S. boats every two weeks. Water conservation was crucial. Food storage was always a limiting factor during patrol. Batteries were recharged by running the diesel engines while the submarine operated on the surface.

During the latter part of the war, the Germans invented the snorkel. This device allowed air to be drawn from the surface into the submerged submarine. The fresh air was required for the diesel engines; crew comfort was a secondary issue. While they were using the snorkel, the submarine was submerged at a very shallow depth, approximately the same as periscope depth. The invention of the snorkel allowed the batteries to be recharged while the submarine was barely submerged. The invention of the snorkel made the submarine less vulnerable to detection and attack than when it operated on the surface.

The "standard" U-boat, which comprised the bulk of the German underwater fleet for most of the war, was highly maneuverable and able to

---

[7] Hughes, Terry, and John Costello, *The Battle of the Atlantic*, Dial Press, 1977.

dive in less than 30 seconds. The success of the Wolf Pack's tactics depended on its low profile, enabling a submarine to take advantage of its 17 knot surface speed to operate as a submersible torpedo boat and penetrate the convoys undetected at night. By 1942 increasing numbers of radar equipped Allied escorts and aircraft were breaking up the Wolf Packs. Admiral (ADM) Karl Dönitz, the commander of the German submarine force, was already planning to replace the standard fleet submarines, and their Wolf Pack tactics, with a new generation of "super" U-boats, designated type XXI submarines. These U-boats were developed to have an underwater speed fast enough to outrun the surface ship escorts. A crash construction program was instituted in the summer of 1943. In spite of the massive industrial effort, however, none of the type XXI boats saw operational service in time to affect the course of the war.[8]

Because of the characteristics of the submarines, the surface attackers had developed some very successful tactics. The principal means of detection were the *Asdic* units.[9] The Asdic sent out a series of sound pulses. If a submarine was located within range (effective range was approximately 2500 yards—or 2.5 kyd or 1.2 nautical miles), then an echo could be heard. This echo was detected by the sonarman (good ears were important). The times between when the pulse was transmitted and the echo was received were an indication of the range of the target. The echo was received from a particular direction.

The use of the two pieces of information, range and direction to the target, allowed weapons to be fired at the target. These weapons were depth charges, and for a depth charge to be lethal, it had to explode within 20 feet of the submarine.[10]

The most effective methods of attacking the submarine were the joint operation of two surface-ship attackers. One ship would attempt to remain in sonar contact with the target, the other would make the actual attack. If the sonar ship came too close to the target submarine, blind spots occurred that made the final weapons release occur in the blind. These blind spots were due to the multiple echos received by the sonar operator, and in essence, the operator lost the ability to distinguish the details of the

---

[8] Ibid.

[9] Asdic was the British term for sonar and was a device for locating submerged submarines by using sound waves. It took its name from the Antisubmarine Detection Investigation Committee that had begun its development in 1917.

[10] Ibid.

submarine echos. By using a series of probable positions of the target (submarine) as the surface ship reduced the range, an estimate could be made where the submarine probably was in the final part of the attack. The surface ship released its depth charges to the probable position of the submarine. Since the speed of the submarine was slow, and the relative speed of the surface ship was high, the probable position of the submarine for the final attack was generally quite reasonable.

As the war progressed, the tactics of the surface-ship attackers became more sophisticated and the escorts became more capable. Since the toll of German submarines steadily mounted and the Battle of the Atlantic was effectively won by the Allies, why was the U.S. Navy worried about submarines immediately after the war and into the 1950s? That question had two answers.

First, the Germans never had more than 50 submarines operating in the Atlantic Ocean at any time during World War II; the average number was probably closer to 20. Of a total daily average fleet of approximately 380 submarines, only about 100 were totally operational. In fact, the daily average number of operating submarines in the entire German Navy never totaled more than 445.[11] With the small number of submarines actually operational, the German submarine force had almost brought starvation to Great Britain and nearly halted the effective prosecution of the war by the Allies.

The Soviet advance from the East allowed them to capture the Baltic area of Germany. This was the location of the German submarine bases. At the end of the war the German submarine officers and men, as well as the new Type XXI submarines had fallen into Soviet hands. The acquisition of the German submarines added to their own submarines, caused the Soviet submarine force to expand to more than 400 operationally ready submarines.

Second, ASW had shown itself to be a difficult game to learn and play. Much of the capability of the U.S. Navy had disappeared with the rapid return to peace after the war, and there was a dearth of experienced people. In effect, the U.S. Navy had lost its hard-won ASW skills, and the acquisition of the type XXI submarines, as well as continued submarine development by the Soviets foretold a very capable opponent.

Aspects of the ASW Problem

The ASW problem has four distinct phases: detection, classification, localization, and kill. Each part of the problem is unique and demands

---

[11] Ibid.

different skills. Detection is the first part of the problem. In the vast ocean it is necessary to locate a submarine. The submarine moves in three dimensions, which makes the problem far more difficult that locating a surface ship, which is restricted to only two dimensions. The submarine cannot be seen in the water, once it is submerged. The surface ship can be seen from airplanes or other surface ships. (Satellite surveillance of the ocean surface was in the far distant future). Detection of the submerged submarine was going to be dependent upon the signals that it transmitted. These were acoustic signals, and they were sent out just due to the operation of the submarine itself.

Once a signal was detected, it was necessary to decide whether the signal came from a submarine, or from a surface ship. It was not always easy to decide what mechanical systems had generated the signal. Submarines and surface ships have similar equipment, so the classification of a received signal was not easy. Also, acoustic signals could travel for very long distances so a signal could come from a ship or submarine that was close or across the ocean.

There was a substantial disparity between the effective detection range of an acoustic signal and the effective range of a weapon. In simple terms, to use a weapon to kill a submarine, the attacker had to be very close to the submarine. But, the platforms that made the initial detection of the possible submarine were very far away. The information that a detection had been made by one type of system had to be transmitted to another type of system that was on a platform that was close to the possible target. The attacker had to re-acquire the target, using its systems, before it could make an attack.

Once detection had been made, and the target had been classified as a submarine, and an attacker had re-acquired the target, it was prepared to launch its weapon to destroy the submarine.

In order for a submarine to be successfully killed, all phases of this process had to be successfully accomplished.

The physics of sound propagation, signal strength, and receiver capability play a role in detecting a signal. Substantial efforts had been made in attempting to detect submarines using acoustics; it is the natural physics of the situation. Water allows sound waves to propagate easily. The distance sound waves travel in the ocean is dependent upon their frequency. Low frequency sound waves travel farther than high frequency sound waves. In the course of their travel, some of the acoustic waves strike the ocean bottom and the ocean surface. Depending upon the angle they strike the boundary, either bottom or surface, as well as their frequency, the amplitude of the sound waves may be attenuated. Sound propagates throughout the oceans for distances that are dependent upon the local boundary conditions[12] and the frequency of the signal.

The controlling parameter is really the signal-to-noise ratio (S/N)[13]. This ratio is the magnitude of the signal strength divided by the magnitude of the total noise, both measured at the receiver. If the signal is much stronger than the noise, it should be easy to recognize the signal. If the signal is much weaker than the noise, then the signal cannot be recognized. This is expressed in the sonar equation.[14] For mathematical reasons, the sonar equation is expressed using logarithms. The signal-to-noise ratio (S/N) is actually a logarithm. (It is the characteristic of logarithms that if the actual ratio is one, than the logarithm is itself, zero.) Thus, if

$$S/N=0$$

it means that the signal strength and the ambient noise are equal. Further mathematics now become involved. While the value of S/N is of interest, the real focus is on our ability to actually make a detection. The probability of making a detection is a function of the value of the signal-to-noise ratio. When S/N=0, the probability of detection is 50%. If the S/N could be increased, then the probability of detection would also increase. The key point is that increasing the S/N is an important part of the solution to the detection problem.

Sound (acoustics) is not the only way to make detections of submarines. A submarine, just like any other signal source has a spectrum of signals. Some of these signals are in the non-acoustic part of the spectrum. For example, the submarine can radiate electromagnetic waves, optical waves, or even heat. Theoretically, all these signals are detectable. The only real issue is, at what range?

Submarines operate in the oceans of the world. In order to find submarines, it is necessary to constantly survey the oceans. The characteristic dimension (distance from one side to the other) of the Atlantic Ocean is about 3000 miles, and the Pacific Ocean is even larger, about 6000 miles. The Mediterranean Sea has a much smaller characteristic dimension, about 300 miles. If we assume that the receiver which can hear

---

[12] *Boundary condition* means the edges of the area being worked in.

[13] See Appendix F, "The Basics," for more detailed information.

[14] Ibid

the submarine is placed in the middle of the region, then it is necessary to hear the submarine at a range of about 1500 miles in the Atlantic, 3000 miles in the Pacific, and 150 miles in the Mediterranean.

There is also a problem with the angle subtended by the submarine to the receiver. If you consider that detection can be made at a range of 200 miles and that the characteristic dimension of the submarine is 300 feet long, then the angle subtended to the receiver is approximately 0.015 degrees—not a large number. That is roughly equivalent to being able to see a fly (1/4 inch in size) on a wall at a distance of 100 feet!

What size target could be seen if the receiver could look in 1-degree segments? For an angle of 1 degree at a range of 200 miles, the target would be approximately 3 miles wide. It is clear that for a receiver to know the precise location of a target, it would be helpful it they were close together. Thus, when a detection is made in a particular ocean region (only about 400 miles across, since the receiver is sitting in the middle of the region), it can't specify the precise location of the target. It can tell another platform that a target is at a particular point, with an accuracy of about 3 miles. The closer platform can then go to the general area of the target and "reacquire" the target. This is known as the "hand-off" problem.

After a detection is made, the next step is to decide whether or not the object is a submarine. During WWII, the principal means of acoustic detection was active sonar. The sonar transmitted an active ping. The detection was made by hearing the ping after it was reflected from the target submarine. The detection was made at ranges generally less than 1 nautical mile. If a detection was made at that range, it was generally a submarine. It must also be admitted that the existence of ocean anomalies, such as current shears, hot spots, reflections from the bottom and surface, and biological reflectors (schools of fish) could, under certain conditions, also generate a reflected signal, but that was relatively infrequent.

During the Cold War the detection was made through passive listening. The source signal was the sounds generated by the submarine. The primary frequency of the signals from other countries' nuclear submarines was generally 50 Hz , which was the frequency of the electrical power-generating equipment and other machinery. Depending upon how this equipment was quieted, through careful design and even more careful methods of attaching the equipment within the submarine, the submarine would generate a certain spectrum of frequencies which were reduced in amplitude. Regardless of the efforts made in quieting, each submarine always has a radiated noise spectrum. The more care made in measuring these spectra, and then in trying to reduce the signatures, the quieter the submarine. The quieting of the submarine was literally a question of life and death, since noisy submarines are easily detected, and detecting quiet submarines is much harder.

An acoustic signal was detected by first listening to bank of filters. These filters are placed in parallel to each other, so the arriving signal was filtered by all the filters, simultaneously. The arriving acoustic signal may consist of many frequencies. It was the function of each filter to remove all of the energy except in that very narrow band that the filter allows to proceed. If there was energy in the band that a particular filter allows to proceed, then it can be displayed. It is then relatively easy to decide whether there was energy at a particular frequency or not.

These filters could be made with amazing accuracy. The bandwidth that could be detected through a filter was substantially less than 1 Hz. Thus, if you were looking for a submarine that had a 50 Hz signal, then you would use a narrow bandwidth filter centered about 50 Hz. Since other signals were associated with a submarine instead of only a 50 Hz signal, you would listen with many filters.

The filters solved the problem of looking at the frequency spectrum of the signal. But the filters did not provide any information about the direction from which a signal arrived. It was also important to try to find out from which direction the signal was arriving. Receiving systems had to be designed, built, and installed that would provide directional information.

A detection would be made with the receiver by looking in a particular direction, and then analyzing the signal using a bank of filters, simultaneously. For a submarine detection to be declared, there had to be at least two lines (on the equipment that was used, a line was drawn on special paper, which was a representation of the specific frequency) at the right frequencies and sometimes more than two. This total signal with its corresponding frequencies was given a name that would characterize the target. If a submarine detection occurred, this name would be sent to other naval platforms and the characteristics of the signal would be carefully noted.

If the detection did not have the right combination of frequencies, then it was considered to be an anomalous detection and was not pursued. The detection was given a name, however, and the signal characteristics were dutifully recorded for possible analysis at another time, and also with possible comparison to previous anomalous signals.

The initial detection almost always was made using the Sound Surveillance System (SOSUS), a system made up of arrays of hydrophones placed on the ocean floor in suitable locations. This system of arrays of hydrophones takes advantage of the natural shape of the ocean bottom.

The output of each hydrophone is sent to a special naval station, located on shore. The distances from the hydrophone arrays to the shore station is hundreds of miles. The hydrophone signals are transmitted over a cable system which has numerous repeaters. The repeaters are designed to amplify the signal as it travels along the cable.

SOSUS

The Sound Surveillance System (SOSUS) is a series of arrays placed on the ocean floor, taking advantage of the ocean bottom shape, cabled to shore with undersea cables. These cables consist of hydrophones arranged in a linear array. Each hydrophone output is sent back over hundreds of miles to a station on land to be processed.

From the advent of the Cold War, the Navy essentially tried to fight the submarine battle on a platform basis. The introduction of SOSUS into the battle created a coastal listening system to detect Soviet diesel submarines off the coast of the United States. When the Soviets introduced nuclear submarines, SOSUS was expanded to cover additional ocean areas. There was success and there was failure, but eventually it became clear that no single platform—whether ship, aircraft, submarine, or SOSUS—could "solve" the problem. It would be necessary to consider integration of many assets to focus on the ASW problem.

Much of the focus on the solution to the ASW problem was on the development and use of special equipment. But, part of the operations of any force is the behavior of the personnel. This fact allowed the development of analysis of the operations of systems. The Navy had used these analysis procedures during WWII. Called Operations Analysis, this technique could provide clues on how to hunt the Soviet submarines. Human beings have a tendency to do things in standard ways. This is reflected in submarine operations. When a Soviet submarine was detected in a particular ocean region, it was reasonably certain that the submarine was going to a particular area and was going to use standard routes to go there. Thus, when an initial detection was made in a particular large ocean area, and it was confirmed to be an enemy submarine, it was considered highly probable that the enemy submarine was going to a place where its predecessors had gone before. This information was well known to the U.S. Navy and had been obtained through trial and error and intensely hard work by thousands of dedicated people over a long period of time.

Once detection and classification were made, the multi-engine aircraft community became players in the game. This community originally flew multi-engine propeller P2s and then P3s of increasingly upgraded capability. These aircraft would fly to various areas that historically were good places to resume the search for the enemy submarine. They would fly various patterns and then drop *sonobuoys,* which are acoustic listening devices that can be dropped into the sea from an aircraft or surface ship, although that was rather rare.

Part of the sonobuoy remains floating at the surface. The rest of the sonobuoy, which is a hydrophone connected to a long wire, falls to a selectable depth, and activates, i.e., it listens. If the sonobuoys hear something, then the signal is transmitted to the aircraft overhead. The signal is analyzed on board the aircraft using special equipment, similar to that equipment found in the SOSUS stations. The characteristics of the analyzed signal are compared to the signal characteristics of the signal that was detected originally at the SOSUS station. If they are the same, then it is considered that a hand-off has successfully occurred. There is reasonable confidence that the SOSUS station that made the initial detection, and the aircraft are listening to the same submarine.

The detection range of the sonobuoy is only a few miles although, as always, anomalies are present under certain circumstances. Generally, however, if a detection is made by the P3 aircraft, then it is reasonably certain that the submarine is close to the sonobuoy, i.e., within a few miles.

Now the situation becomes quite interesting. The aircraft on station (flying above the area in the vicinity of the submarine) attempts to establish a track of the submarine. The aircraft drops additional sonobuoys, and hopefully makes additional detections. This allows the establishment of a probable submarine course and speed. Once there is reasonable certainty in the course and speed of the submarine, the last phase of the game begins. This requires another hand-off, but this time to a U.S. attack submarine.

Communications ensue that provide all the information about the enemy submarine to the United States submarines operations center in the particular ocean. An attack submarine, near to the probable location of the Soviet submarine, is informed about the track of the enemy submarine and is given the characteristics of its acoustic signal. The U.S. submarine then uses its sonar and searches in the area that the enemy submarine should enter if it follows its probable track. Once detection is made, the U.S. submarine assumes a position behind, above or below, and to the side of the enemy submarine and continues to monitor the acoustic signals of the enemy submarine. This information, which is sensed by several different systems on board the U.S. submarine, is analyzed and provided to a computer that programs the information into the torpedo that would be used to kill the enemy submarine. This information provided to the torpedo is known as the *firing solution*.

At this point the game is over. The U.S. submarine trails the Soviet (or other enemy submarine) for as long as required, generally a period of days or weeks. A firing solution is maintained at all times. If a state of hostilities should be declared for any reason, then the U.S. submarine would be contacted with the appropriate communications equipment and ordered to destroy the enemy submarine that it is trailing.

This entire process was referred to as a game in the sense that it was a continual state of practice for the U.S. Navy. Certain important issues have arisen in the course of this discussion. All the phases of the game—detection, classification, localization, and kill—required complex and different equipment, as well as a large number of players. The communication problems were manifold and world wide. Timing was essential for success. Close and careful coordination had to occur between many different platforms. Each different class of platform communicated with their operational center. The different operational centers were frequently located hundreds, even thousands, of miles apart. They communicated with each other, and coordinated the activities of their own platforms. Each platform and operational center also had different cultures—not an insignificant problem. Sometimes the differences in cultures caused misunderstandings between the different platforms and their operators. Each player in the game tried to increase their skills and capabilities. It was clear that enhancements to any part of the process contributed substantially to the probability of success of the combined efforts.

The Technical Problem

The technical problem is the application of the passive sonar equation. The sonar equation can be written as

$$S/N=SL-TL-AN+DI.$$

where S/N is the signal-to-noise ratio, SL is the source level, TL is the transmission loss, AN is the ambient noise, and DI is the directivity index of the receiver. For a 50% probability of detection, S/N=0. Thus, in that case,

$$TL= SL- AN+DI.$$

For a given ocean and season and for a given geometry (source and receiver depth specified), the TL can be calculated as a function of range. The SL is a function of the target (enemy submarine) and is not under our control. But, it can be characterized by a parametric set of values that should include any reasonable level. The AN is calculated from a knowledge of all the positions of the surface ships on the ocean in question, as well as their source levels. The DI is calculated, and depends upon the characteristics of the receiver. Thus, after all that calculation, a range can be found for the specific parameters. Since the AN and TL are azimuthally dependent, the probable detection range is also a function of azimuth. It is easy to see that the number of calculations that can be made is large. In fact, as each term in the equation becomes more complicated, the number of calculations grows algebraically.

As an example, consider a case where the submarine can be either deep or shallow, as well as the receiver. Consider three different frequencies, and also consider that the azimuth of 360 degrees is to be considered in 5-degree arcs. That means we must consider 72 different sectors. So far we need to calculate the sonar equation with 2x2x72=288 separate runs. (It turns out that we can apply frequency corrections without recalculating!) This is for a simple ocean with range independent velocity profiles: one velocity profile is all we need. Further, we will consider the shape of the bottom, i.e., the bathymetry, to be flat: another simplification. Thus, all we have to do is calculate the terms of the sonar equation 288 times. This calculation is hard to do for humans but is exactly what computers are designed to do. Now, we have to take into account the actual receiver characteristics, which we can do by convolving the receiver shape with the appropriate values. The inference to draw from this small explanation is that many calculations are required to predict the performance of a receiver attempting to detect a submarine.

Let's return to the sonar equation. All four terms in the equation depend on physics. One term, and one term only, however, is almost entirely under our control: the receiver. We can't do anything about the physics of propagation; we can't control the source level of the target submarine; and we certainly can't control the ambient noise. But, we can design receivers in a variety of different geometries with any number of hydrophones that we desire. Further, we have at our complete discretion the depth and the location where we place the receiver.

We would like to have a receiver that will allow us to look in a particular direction and to eliminate all the spurious signals (noise) arriving from other directions. This look direction should be specified arbitrarily. We would like the beam width, associated with the beam looking at the angle we desire, to be very narrow. So, mathematically, we would like to have very narrow beams and very low sidelobe levels. In fact, when and if we encounter specific noise sources, we would like to point *nulls* in their direction (nulls are directions in the receiving beam pattern that receive either little or no energy), which is entirely possible.

The device that we need to build is a linear array of hydrophones. In fact, without getting into the mathematics, we can build an array that is very long and has so many hydrophones that we can do everything that is required. This long array can be placed at an arbitrary depth in the water and is towed behind a slow-moving ship. Towed arrays have been built and used that are over 3 nautical miles long! Shorter arrays with a smaller number of elements have “fatter” beams and possibly higher sidelobe levels.

Conceptually, the process is then quite simple. After making all the calculations of the first three terms of the sonar equation, the receiver is placed in position, looking in a particular direction that will enhance your probability of detecting the enemy submarine, or for the given situation, increase the distance at which the probability of detection of the submarine will be 50%.

These calculations are prodigious. They take substantially sophisticated computer models that have been tested against known experimental results. Not only do the models have to be built; they must be verified by comparing their predictions to measured data from substantial controlled experiments at sea using submarines, acoustic sources, and several types of measurement equipment.

Once the data have been taken at substantial cost in time, money, and effort, the measured results must be carefully compared to the results predicted by the models. When a discrepancy occurs, it must be explained and understood and then the models must be corrected. This process of measurement and comparison to model outputs is called *verification and validation* of the models. Since the models are the only tools available to predict what will occur in a specific situation, confidence in the model results must be generated. This process is more than an academic or scientific exercise. These models predict the acoustic results of the assumed scenario. The distribution of forces at sea, and people’s lives depend upon the model’s ability to predict what will occur.

The Long Range Acoustic Propagation Project (LRAPP) spent much of its effort building the models, taking the at-sea measurements, and then employing the verified and validated models in actual operations to see if they could solve the ASW problem.

CHAPTER Four
Advent of the Cold War

General Comments

The advent of the Cold War brought forth the possibility of the United States being attacked on its own shores. These attacks could be launched by submarines that carried missiles. The United States government felt that It was imperative to establish defenses that would warn of the presence of the Soviet submarines near the United States homeland, and also to develop the methods to kill them if necessary.

The success of the aircraft carriers during the Pacific War showed that the United States had the ability to project power anywhere on the face of the planet. Their existence and the power that they could project made them obvious targets for submarine-launched torpedoes and missiles. The ASW problem, then, became two separate problems: (1) dealing with the missile-carrying submarines off the coasts of the United States and (2) the protection of the US aircraft carriers that operate throughout the world's oceans from attack submarines.

Missile Submarines

The original missile submarines used by the Soviets were extensions of the World War II submarines. They were diesel powered and battery driven. They were large for the time but were tiny compared to the ballistic missile submarines of today. They had to surface to fire their missiles and, even with haste, it took some time for the missiles to be launched. The issue was, could the United States detect these submarines? It was certainly believed that if the submarines could be detected, they could be attacked and destroyed. The U.S. Navy decided that it could detect the enemy submarines.

A special project called Jezebel was begun during the mid to late 1950s.[15] Hydrophones were placed in arrays along the coast of the United

[15] Several histories have been and are being written about the SOSUS, which was another name for this project.

States. Then, the hydrophones were connected in a long line, and the signals were transmitted through underwater cable back to shore stations. This process required many skills and techniques:

- skill in underwater cable technology;
- expertise to place the hydrophones in the proper position;
- ability to recover, repair, and replace the arrays and cable when problems occurred;
- expertise to analyze the received signals at the shore stations.

AT&T was the only organization within the United States that had the proficiency and experience to perform these tasks. After some negotiations, AT&T became the sole organization to develop construct, install, maintain, and operate the system under the command of the U.S. Navy. The product was provided by both Bell Laboratories (science and technology, and engineering design) and Western Electric Company (surveys, ocean engineering, and installation). At that time, there was neither expertise within the U.S. Navy to perform such a task nor the time to develop it. The Soviets were already taking a highly aggressive stance throughout the world. The Cold War had begun in earnest.

The initial cost for developing the SOSUS system was staggering. Over $1 billion was budgeted for the system; this amount is in 1950 dollars. There was substantial argument about the expense. Several alternative methods were proposed, but none of them could provide the early detection that SOSUS could—if it worked.

As an example, the cost for a repeater in the transmission cable from the hydrophones back to the beach was $500,000 at that time. The quality of the signals and the techniques of cable manufacturing required that a repeater be placed approximately every 20 miles. It was with hesitation that Western Electric was allowed to begin building and installing these very expensive units.

Exactly how good these units were was discovered and confirmed in the late 1980s. As the SOSUS system was managed and rearranged, different cables and their repeaters were recovered and tested. *After being on the bottom of the ocean for close to 40 years, the repeaters still continued to function, meeting the specifications to within 0.5 dB.* This effort is an astonishing testament to the quality of engineering that was provided to the U.S. Navy by AT&T and Western Electric.

While the cost of the repeaters was high, it was less than the cost to go to sea. The cost for "ship-days," i.e., the cost to operate a cable ship in the late 1980s was $50,000 per day. The cost of recovering and refurbishing the repeaters was substantially greater than the original investment. Having a system that worked without problems for four decades was an amazing feat of engineering and cost effective.

Unfortunately, while the concept of detecting submarines by using their radiated noise was good, it didn't work. Surfaced submarines emitted sound that almost immediately propagated into the bottom. In certain special circumstances, the sound propagated in a channel at the surface of the ocean (*surface channel propagation*). The noises emitted by the post-World War II submarines were generally *broadband* (covered a portion of the acoustic spectrum and were not at a single frequency). That meant that even if a detection could be made, the ambient noise of the ocean itself sounded just like the noise emitted by the submarine; thus, it was extremely difficult to distinguish the submarine's noise from the ambient noise. A submarine operating on diesels, sounds exactly like a surface ship; there was no distinguishing signature. A submerged submarine operating on batteries generated almost no noise, so it was not possible to detect submerged submarines.

It appeared that the advances made by the scientists at AT&T, the remnants of the Harvard group, and other scientists rising within the U.S. Navy laboratories at New London, Connecticut, San Diego, California, and elsewhere were not equal to the task of detecting the Soviet missile submarines of the mid 1950s through the early 1960s. This problem weighed heavily on the minds of the senior leadership of the U.S. Navy.

Attack Submarines

The technical problems associated with the detection and prosecution of attack submarines were the same as those for missile launching submarines. The mission of the United States Navy was to locate and attack both attack and missile submarines, but their capability to do so was limited.[16] During the 1950s the Navy formed hunter-killer groups whose primary mission was to find enemy submarines. The hunter-killer groups consisted of surface ships— destroyers and small aircraft carriers (the jeep carriers of World War II days). The hunter-killer groups had hull-mounted

---

[16] Interview with Captain Robert Adler, USN (Ret.) and Captain Fred Kelley, USN (Ret.).

sonars, designated the SQS-23 and SQS-26 sonars.[17] They spent substantial amounts of time and effort cruising off the coasts of the United States. Unfortunately, their success was minimal.

In addition, several other issues should be noted, although they were not considered relevant then. The ships had no modeling capability on board. Computers, in particular small portable ship board computers, did not exist. There was no predictive capability for the operation of the sonars, as poor as they were, and there was no understanding of the role played by the environment. Later, these issues became critical.

The SQS-26 sonars were specifically built to extend the detection range of the surface ship. The environment was sufficiently understood to recognize that the ability of the SQS-26 sonar to detect actively submerged submarines was sharply curtailed for some unknown reason when the ranges became too large. The process was that the surface ship would detect the submarine using active sonar. Then, as the range increased between the two ships, there was a sharp, immediate loss of signal. It was not a general degradation, but at one moment the signal was there; the next moment it was gone. Clearly, something was happening that not many scientists really understood. Those who did, however, pointed out that the environment—specifically, the sound velocity profile— caused the sound to refract in the ocean in such a way that the reflected signal did not return to the receiver on the surface ship.

After some heated and detailed arguments, it was decided to bounce the tremendous acoustic energy transmitted by the SQS-26 sonar off the bottom. Conceptually, the sonar would send out a signal that went almost straight down to the bottom, perhaps 2 nautical miles away, reflect, and come up and hit the submarine. The reflection of the acoustic energy would take the same path back to the receiver.

Several questions arose immediately: What were the characteristics of the bottom? What was the shape of the bottom? Did it have varying physical properties? Would the sound reflect without loss, or be completely absorbed? Was the reflectivity of the bottom frequency dependent? The SQS-26 sonar operated at 3.5 kHz. The frequency was not variable, but the wave form was. Was the reflectivity of the bottom dependent upon the shape of the wave form?

---

[17] The variable depth sonar did not exist at that time, and the SQS-26 sonar did not come on line until the late 1960s.

No one knew the answers to these questions. The result of the discussions about these questions was the establishment of a world-wide measurement program whose mission was to determine the characteristics of the bottom throughout the oceans of the world, or at least in the Northern Hemisphere. Although this data base[18] was used successfully 25 years later in another context, unfortunately, it never helped to increase the success of the SQS-26 in the bottom-bounce mode.

One other thing was noticed. When the SQS-26 was passive, sometimes a strong contact occurred that the sonarman would incorrectly identify as a signal. Neither a submarine nor surface ships were present in the area of the signal. From the destroyer, the horizon is about 15 miles, and it was evident that no other source of signals was present. Eventually, people decided that these signals were merely anomalies. They were wrong, but it took quite some time to correct their erroneous conclusion (to be discussed later).

Essentially, the aircraft carrier had very little protection. The fleets were arranged in such a way that the submarine would have to penetrate the destroyer screens, which used active sonar to look for enemy submarines in the area. The hope was that the submarine would be detected before it could fire torpedoes at the carrier and, further, that the carrier would be effectively screened by the destroyers and cruisers.

The situation was bleak. The U.S. Navy had been making substantial advances with its own submarine force, and the first nuclear submarine, the USS *Nautilus*,[19] designated SSN571, was "underway on nuclear power," on 17 January 1955. The Soviet Union also developed nuclear submarines, and the first Soviet submarine was underway in the late 1950s. This key event was the salvation of the SOSUS system.[20]

---

[18] The Alpine Geophysical Data Base was compiled for NAVOCEANO.

[19] http://www.ussnautilus.org/history.htm.

[20] Interview with Mr. Wendell Sykes.

CHAPTER Five
Fortunate Events

General Comments

Soviet nuclear submarines had appeared on the scene. They not only had turbo-generators that produced electricity at a nice, reassuring 50 Hz, but they also were very noisy. Their engineering was adequate but did not focus on the myriad details that quieting demands. In effect, the Soviets made submarines that worked but were easily detectable. This flaw was the salvation of the SOSUS system. U.S. submarines were much quieter, but they also were relatively easy to detect. This topic soon became a policy issue. The U.S. Navy decided that U.S. boats could not be detected and that Soviet vessels could be detected, when in fact, both could be detected quite easily in the early days (early 1960s). However, only Soviet boats were reported as being detectable.

The U.S. Navy remained concerned about ASW. As a direct result, a new organization, the Director of

Vice-Admiral Charles B. Martell

VADM Martell's career did not point him toward ASW, but he followed a career that frequently led him into direct contact with the scientific community. His career was extraordinarily successful. In 1930, he graduated 15th in a class of 405 from the U.S. Naval Academy. He served in many positions within the surface Navy. He was awarded the Legion of Merit (twice) and the Distinguished Service Medal. While serving as Deputy Director of Defense and Engineering (Administration and Management), he was promoted to Vice Admiral (1 March 1961). In August 1963 he became the Commander of the U.S. Navy's Second Fleet. The following May, he became the Director of Antisubmarine Warfare Programs. It was for his service in that position that he was awarded the Distinguished Service Medal. On 1 November 1967, he was transferred to the U.S. Navy Retired List. *(Biography courtesy of the U.S. Naval History Division, U.S. Navy Operational Archives)*

Antisubmarine Warfare Programs— designated OP-095—was formed as part of the Office of the Chief of Naval Operations (CNO) in 1964 and, as mentioned previously, its first commanding officer was VADM Martell. OP-095's mission was to focus on the ASW problem.

As always in a bureaucracy, there was a question about the reporting chain. The question was whether OP-095 should report to the Assistant Secretary of the Navy for Research and Development (ASN(R&D)) or to the Chief of Naval Operations (CNO), who is the military head of the Navy. After some deliberation, VADM Martell decided that OP-095 would report to the CNO, but that he would have direct access to the ASN(R&D). His decision was extremely important, since as OP-095, he was the czar of ASW for both the Material and Operations sides of the U.S. Navy. Reporting directly to the CNO, he had both the responsibilities and the power for setting requirements within the designated area for the CNO. However, the financial resources to fund the program were dependent upon ASN(R&D).

In a single, brilliant stroke, VADM Martell had direct access to both the requirements and the financial chains of command with the U.S. Navy. Further, he clearly recognized from the beginning of his tour that the ASW problem could be solved only by the thorough integration of science and naval operations. He was in an excellent position to command the naval part of that partnership; the problem, then, was how the scientific community would respond to the challenge, and how it could be integrated into the picture.

At the same time another event occurred. Dr. Art Maxwell, formerly the Provost at the Woods Hole Oceanographic Institution, resigned from his position as head of the Ocean Sciences Group (designed Code 408 on the organization chart) at ONR. The Chief of Naval Research (CNR) required that someone immediately fill this position. Dr. J. Brackett Hersey, also formerly part of the staff at Woods Hole, was selected for the position as Code 408. The placement of Dr. Hersey as Code 408 was extremely fortunate, since he had excellent relations with the academic scientific community, rapidly earned the trust of senior naval officials, and was very politically astute.

At that time, as today, there was an Oceanographer of the Navy, who was in charge of all issues related to the oceans. He had a Deputy Oceanographer on the staff. The Oceanographer reports directly to

ASN(R&D). This chain of command is unusual, since most flag officers[21] report through the CNO. The Oceanographer was different, however, primarily because of the strong emphasis on providing scientific information, as well as operational information that was heavily based upon science.

The Oceanographer of the Navy and the Chief of Naval Research also reported directly to ASN(R&D). The Deputy Oceanographer, who was part of the Office of the Oceanographer, simultaneously filled the position of Code 408 in ONR. He was "double-hatted.[22] As part of the administrative process, Dr. Hersey was associated with both the Oceanographer and the CNR. Because of his long association with Woods Hole and the oceanographic community, he had an excellent reputation within the academic community and a clear understanding of the scientific and technical resources that could be brought to bear on a naval problem.

At that time, the position of ASN(R&D) was filled by Dr. Robert Frosch, a former director of Hudson Laboratories, (associated with Columbia University) and completely familiar with the problems of underwater acoustics.[23] As ASN(R&D) he had substantial power. He was heavily involved in dealing with the major funding issues that are characteristic of Washington operations. His past relationship with Dr. Hersey and his close relationship with VADM Martell allowed him to leave the detailed operations

---

[21] Flag officer: When someone becomes an admiral, he or she is given a flag with their number of stars on it, i.e., a two-star admiral's flag has two white stars on a blue background. This flag is displayed when the admiral comes to a ship; the admiral's aide provides the flag to the ship before his or her arrival. When the flag is raised, everyone knows that the admiral is on board.

[22] Double-hatted means that a person officially performs two jobs at the same time. A single person can report to two entirely different chains of command. While such a designation is frequently difficult and cumbersome to implement, sometimes it provides substantial power in a particular individual. For example, in the case where operational command authority is one chain of command and financial responsibility and authority is another chain of command, there is frequent tension and disagreement between the two chains. But when a single person is double-hatted and is a member of both chains, it provides substantial opportunity for accomplishing both missions effectively.

[23] Interview with Dr. Frosch.

to them. He also knew that if a scientific issue of great importance arose, Dr. Hersey would inform him. He further knew that if important naval operations issues arose, VADM Martell would inform him. In effect, due his personal confidence in both VADM Martell and Dr. Hersey, both OP-095 and ONR Code 408 had great latitude in taking the steps they thought were necessary to address and solve the ASW problem.

In addition to being part of the reporting chain for the Oceanographer of the Navy and the CNR, the ASN(R&D) also controlled the funds for research, development, test, and evaluation (RDT&E). Other types of funds were controlled within the Operations, Navy (OPNAV) and by the Chief of Naval Material, (CNM).

Another point should be mentioned: OPNAV dealt with requirements, and the system commands (SYSCOMS) dealt with actual material. The SYSCOMS, reported directly to the Chief of Naval Material, who had the money and the wherewithal to design, build, deliver, and maintain the equipment that was used by the fleet. The amounts of money that flow through the SYSCOMS, even in 1965, were in the billions of dollars.[24] The budgetary battles that were fought both within and between those organizations were fierce and brutal. There is always limited funds. The struggle for funds is waged internal to the Navy, and then between the services within the Department of Defense. To continue to be funded at the planned level it is necessary to regularly justify the program to the senior executives throughout the year, and in particular during the formulation of the annual budget. Success in protecting or even enhancing the funding of a program directly reflects upon the skills of the Program Manager. Career enhancement of the Program Manager is directly related to success in the programmatic and budget process. Failure to defend the program can mean either early retirement or removal from the ranks of probable promotion candidates.

At that time, the Secretary of Defense was Robert S. McNamara. During the period 1961-1968 he completely changed the system by which funds and programs were supported in the Department of Defense. The McNamara Planning, Programming, and Budgeting System not only encouraged adversarial combat between the systems and program offices; it demanded it. The McNamara System was based upon a belief in the Darwinian concept of natural selection. Sometimes that concept worked but

---

[24] The SYSCOMS were the Naval Electronic System Command (NAVELEX), the Naval Ship Systems Command (NAVSEA) and the Naval Air Systems Command (NAVAIR).

not always. Naval careers were dependent upon the proper handling of program schedules and finances; while it was certainly hoped that good and valuable systems would spring from such struggles, that was not always the case. No naval officer was ever rewarded for pointing out that his program was not as good as a program supporting a competing system. This fact of life caused substantial competition between programs for limited financial resources.

One other fortunate event was that the Oceanographer of the Navy was also responsible for the Fleet Numerical Weather Central (FNWC),[25] located in Monterey, California, on the grounds of the U.S. Naval Postgraduate School. The commanding officer at that time was Captain (CAPT) Paul Wolff, who had a substantial interest in underwater acoustics.

FNWC's responsibility was to provide minimum distance routing for all the ships and aircraft of the U.S. Navy. Their mission was ostensibly trivial, since the minimum geographic distance between any two points on the surface of a sphere (the earth) is known as a great-circle route. Ships and aircraft use fuel, and the weather causes greater use of fuel if conditions are not optimum. Thus, FNWC was tasked to provide optimum routing for all ships, given time-dependent weather conditions, to minimize fuel usage, as well as wear and tear on their equipment and personnel. This mission was very important, and the solution

Philosophy: Captain Paul Wolff

CAPT Wolff's place in the chain of command was through the U.S. Navy Weather Service. He had a substantial interest in underwater acoustics and felt that it was the Weather Service's responsibility to provide performance predictions for the ASW systems to fleet units throughout the world. This capability required substantial computing capacity, as well as a world-wide rapid, reliable communications system.

[25] FNWC was also known as the Fleet Numerical Weather Facility (1966) and the Fleet Numerical Oceanography Center (1970), and by the nickname "Fleet Numbers." Although still located in Monterey, FNWC is now a field office of the Naval Research Laboratory (NRL), Washington, D.C. For clarity and consistency, however, this fine facility will be referred to as FNWC throughout this document.

to the problem to providing the needed information to the fleet in a timely manner demanded the best technical minds in the nation.

Successfully performing the mission meant that CAPT Wolff's staff of oceanographers and meteorologists had to develop computer models, the ability to run them on the best computers of the day, and communicate world-wide with the most up-to-date communication system possible. FNWC developed a computing capability second to none in the Free World, as well as a communication system that operated in parallel to the official U.S. Navy communication system.

The specialists at FNWC were divided into two groups of subspecialists: the meteorologists (weather guessers or balloon blowers[26]) and the oceanographers, who were further subdivided into groups of physical oceanographers and underwater acousticians. Fulfilling FNWC's principal mission involved the meteorologists, but CAPT Wolff had plans for the oceanographers. However, he did not have connections to organizations with an ASW mission or to organizations with the scientific knowledge and financial resources to attack the problem.

The formation of OP-095, the arrival of Dr. Hersey, and their connection to CAPT Wolff was serendipity and of great value to the U.S. Navy, the United States, and the citizens of the Free World.

Fundamental Concepts

When the U.S. Navy finally decided that they needed OP-095 to directly focus on ASW, this move was more than extraordinary within the OPNAV organization—it was unique. All of the organizations within OPNAV have historically been organized along platform lines, i.e., submarines, surface ships, and aircraft. The naval officers had their specialities as submariners (bubbleheads), or surface ship officers (blackshoes), or aviators (brownshoes). These officers were focused only on the issues associated with their platforms. Before the formation of OP-095, no organization within OPNAV focused on the *function* of Antisubmarine Warfare.

For whatever reasons, the concept of selecting a czar for ASW was brilliant and orthogonal to standard thinking. Using the jargon of today, it was "out-of-the-box" thinking. The selection of VADM Martell as the first OP-095 was extraordinarily successful.

---

[26]This term was due to the weather balloons launched freqeuently by the meteorologists.

Now, the proper men were in the proper positions at the right time. The administrative system also had the right organization to allow their components to effectively deal with one another. OP-095 reported to CNO and had funding ties to ASN(R&D). Both the Oceanographer of the Navy and the CNR reported to ASN(R&D). Dr. Hersey was double-hatted as the Deputy Oceanographer of the Navy and as Code 408 within ONR. He came from the academic underwater acoustics community, and was a friend and trusted confidante of Dr. Frosch. CAPT Wolff at FNWC also reported to the Oceanographer of the Navy.

It was clear to VADM Martell that if there was an operational solution to the ASW problem, science and the scientific community would provide the answers.

Money Designations[27]

A brief explanation of the method in which funding is distributed and employed is valuable for understanding the uniqueness of the establishment and placement of LRAPP within ONR.

The designations of different types of money and its flow within the U.S. Navy are in a constant state of flux. It is simply beyond the capabilities of the author to provide these details. The history of the changes of power within the U.S. Navy are complex and arcane. From 1960 to 1990, the changes were substantial and continual; even if the author could provide the correct description of the process, it would certainly have to be altered several times to accommodate the immense changes that occurred during this time.

Designation of Research and Development Money within the U. S. Navy

The Research and Development funding is referred to as *category 6 money*. The steps for the development of a system are divided into six separate categories and have numeric designations.

- *6.1 funds* (Basic Research) are used to develop the knowledge base.
- *6.2 funds* (Exploratory Development) are used for the development of technology and examination of concept feasibility.
- *6.3 funds* (Advanced Development) are used to determine concept feasibility and to demonstrate system and subsystem feasibility.

---

[27] From RDT&E Management Guide (circa 1975).

- *6.4 funds* (Engineering Development) are used to further develop uses for advanced systems already developed.
- *6.5 funds* (Support Functions) are used to provide management and support to the administration of the RDT&E process and programs.
- *6.6 funds* (Systems Preparation) are used to complete the preparation of systems for Fleet use.

Generally, the Technology Base category is supported by both 6.1 and 6.2 funds. The application of science and technology in support of system development integrates 6.1, 6.2, and 6.3A funds. The 6.3A funds are early 6.3 funds. The Test and Evaluation of the formerly conceptual, now preliminary working systems is supported by 6.3B, 6.4, and 6.6 funds.

APN:	Aviation Procurement, Navy
O&M,N:	Operations and Maintenance, Navy
OPN:	Other Procurement, Navy
R&D:	Research and Development
SCN:	Ship Construction, Navy

Broadly speaking, and without too much detail, the funding allocated to the 6.1, 6.2, and 6.3A categories falls under the control of the ASN(R&D) and the CNR, who, by law, is the mission sponsor for 6.1 Navy funds. ONR had also historically been given responsibility for managing a few 6.2 funds. After the disestablishment of NAVMAT, ONR became responsible for all 6.2 funds. The funds for the 6.3B, 6.4, and 6.6 categories fall under the control of designated Program Managers, but the ultimate decisions are made in the SYSCOMs. The types of available money are manifold and are colloquially referred to as the "color of money." Some examples of the different types are R&D, SCN, APN, OPN, and O&M,N.

Once funds have been designated a certain "color," e.g., R&D category 6 funds, they cannot be changed without Congressional approval. Further, within a specific category (for example, 6.2) funds cannot be transferred to another category, even within the R&D designation, without Congressional approval if they are above the threshold of $4 million. Of course, there are always exceptions and financial limits, but in essence, funds that are designated through the budget process as 6.2 Exploratory Development R&D funds must remain so.

The small keyhole that allows some latitude with funding is deciding whether or not a project category is 6.1, 6.2, or 6.3. As stated, 6.3 funds cannot support 6.1 or 6.2 tasks. However, contention frequently arises. For example, assume that a 6.3 program uses 6.3 funds to support a 6.3 task. A problem suddenly arises and the solution, by most measures, would be a 6.2 task. The question is, can 6.3 program funds be used to support that 6.2 task? The answer is a qualified, and highly hesitant, sometimes. It requires a judgment call. The amount of money is always in question; the percentage of the program funds is considered; the intent of the original legislation and somewhat arbitrary divisions between different types of monies and tasks is carefully taken into account. In practice, 6.3 funds are used to support 6.3 programs that have traces of 6.1 and 6.2 tasks.

The dissemination and control of funds within the Department of Defense, DoD, and the U.S. Navy are complex and sophisticated. The key point, as far as this tale is concerned, is that the Office of Naval Research and its commanding officer, the Chief of Naval Research, really controlled all 6.1, 6.2, and some 6.3 funds. The 6.4 and 6.6 funds were always controlled by the SYSCOMs.

CHAPTER Six
LRAPP-Beginnings

General Comments

When VADM Martell and Dr. Frosch organized and sponsored LRAPP in 1967, the program was placed in ONR. At that time, ONR used Basic Research (6.1) funds almost exclusively. Exploratory Development (6.2) funds were also available to ONR, but not Advanced Development (6.3) funds. The placement of LRAPP in the ONR administration, and its associated 6.3 funds, was unique. (It is not so today, but that is a story for another time.) The LRAPP funds filtered down from ASN(R&D) into ONR. The people within ONR were unfamiliar with 6.3 funds that support Advanced Development. The contract administrators were used to operating with the 6.1 and 6.2 cultures, and their types of projects. The objectives and projects of the 6.3 community were sufficiently different from 6.1 and 6.2 programs that it caused some difficulties within the contracting community within ONR. These were eventually all handled, but it was a struggle for all involved.

At that time, the recipients of 6.1 funding were almost exclu-sively universities. However, some government laboratories also received some 6.1 funding. All organizations that received 6.1 funding performed basic research and never felt a sense of urgency to deliver their products; that was the correct way to feel. Their work served as the foundation for naval technical systems and programs that would be executed 20 years in the future. They had no role in trying to stand shoulder to shoulder with the operational Navy.

The recipients of 6.2 funding were not populated primarily by universities. Many Navy laboratories and a few industrial organizations used 6.2 funds. These organizations also were not standing shoulder to shoulder with the operational Navy, but they were a little closer than the basic researchers. The 6.2 culture also felt no sense of urgency in dealing with the existing ASW threat, and that, too, was correct; it was not their job.

LRAPP in Washington

The LRAPP office was co-located with the Maury Center, which was named for LT Matthew Fontaine Maury. The Maury Center was located in Building 58 on the campus of the Naval Research Laboratory.

Dr. Hersey was concerned with the ability of the sci-entific community, as distinguished from the Navy's scientific community, to feel welcome at the Maury Center. This concern led to dispensing with the common security controls associated with all Navy installations. An ingenious scheme was generated. The wire fences that surrounded NRL were extended on large swinging gates. During the day the gates were locked in one position, such that the entrance to the Maury Center was directly accessible to the outside world, with no Navy security controls. At night, the gates were swung into different positions, so that Building 58 and the Maury Center became part of the NRL compound.

Initially, this gave rise to some strange events. If a visitor came to the Maury Center in the late afternoon and held extended discussions with Dr. Hersey and others, they would find that they were suddenly inside NRL. When they attempted to leave, the security guards would politely ask for their badge or pass, which they were required to have whenever they visited any part of NRL. This situation caused some interesting bureaucratic arguments until the security guards understood the unique position of the Maury Center.

Of course, the general occupants also remem-bered to check the time and urge their visitors to leave before the change of the gates from day to night positions.

The recipients of 6.3 funding were basically parts of the Navy laboratory system and industrial contractors. These organizations were much closer to the oper-ational Navy; frequently, some of their advanced development systems were used on a "not-to-interfere" basis in Fleet exercises. The Navy was interested in ensuring that the systems in the advanced development stage could perform the task for which they had been designed. After all, the systems would eventually have to operate accurately in the real world. But this culture was alien to the 6.1 R&D communities that were the prevalent organizations reporting to and supported by ONR. Clearly, the introduction of LRAPP into the ONR community would induce certain administrative strains.

Although VADM Martell had established LRAPP as the only 6.3 program in ONR, it was VADM Edward Waller's opinion[28] that this unusual arrangement was the only way that LRAPP could have survived and prospered. As a Program Manager, VADM Waller understood that LRAPP, as an R&D program , would have been at the mercy of the engineering development programs within the Navy's SYSCOMs,. As the only 6.3 program in ONR, there was no possibility for any reprogramming into other ONR programs. Remember, it is nearly impossible to change money colors, i.e., 6.2 Exploratory Development money cannot be used for Advanced Development programs. The reverse is also true. Therefore, other programs in ONR, which were 6.1 and 6.2, could not use LRAPP money.

> **What is "Not-to-Interfere"?**
>
> "Not-to-interfere" means that scientists are allowed to bring equipment and personnel aboard Navy ships, but could not interfere with Navy operational exercises. For example, a part of shipboard data gathering is to drop dye markers in the water, but scientists were not allowed to do so if their task interfered with any Navy operation.

ONR is unique within the Navy. It is a R&D command and not a platform focused SYSCOM. Because of this uniqueness, there was considerable concern within ONR that LRAPP, as a 6.3 program, was outside the bounds of the ONR Charter—to develop information that will help the future fleets. VADM Waller felt that LRAPP with its developments was doing exactly that: supporting development of future tools focused on the ASW problem that would support future fleet operations.

Another unique characteristic of LRAPP was the inte-gration of scientific exercises with fleet exercises. VADM Waller clearly remembers that he never heard any com-plaints. The fleet commanders were grateful for any help that LRAPP could provide.

---

[28] Interview with VADM Edward Waller, USN (Ret.)

The LRAPP philosophy under Dr. Hersey (the first program manager) was well balanced and well integrated. It was clear from the beginning that (1) the real focus of support to the operating fleet would have to be the extended use of computer models; (2) these models would have to be developed by the scientific community and validated with the results of oceanographic exercises; and (3) these models would have to be run and the results delivered to the operating fleet in a timely manner. These three requirements essentially defined the operations of LRAPP for its entire life.

The computer models were developed by individuals from universities, Navy laboratories, and industry. The oceanographic data, which was both difficult and expensive to obtain, would require assets (ships and associated support equipment) from both the university oceanographic fleet and the Navy laboratory fleet, as well as support from actual operational Navy fleet assets (extremely difficult to obtain). These oceanographic data collection efforts would be supported by people from the seagoing measurement community, generally from universities and Navy laboratories, with some industry participation. The reduction of the massive amounts of data would be done by specialists in such efforts. These specialists came from diverse places, but generally from Navy laboratories and industry.

Dr. Hersey and LRAPP

LRAPP's staff was small. Dr. Hersey had three support people: Mr. Al White, Mr. Ferd Diemer, and Mr. Bert Thompson. Throughout the entire life of LRAPP (approximately from 1967 to 1989), the staff was never appreciably larger because most of the people associated with LRAPP were frequently temporarily appointed. Their billets were in their home duty stations, and they were given temporary duty to support LRAPP. There was always a military officer and two or three secretaries. During the 1970s, LRAPP had nine civilian and one military billet. During the 1980s, this increased to 12 billets.

Dr. Hersey ran LRAPP as he would have run an academic program; after all, his background was from the academic world. He had a clear idea what had to be done, but he did not have the temperament to run a program that was

The fleet computer models operated 24 hours a day, with naval communication capabilities that reached all the corners of the world. This fleet support operation would be within the control of the U.S. Navy. And there was a place for this to happen: FNWC, under the command of CAPT Paul Wolff.

The LRAPP Exercises

Dr. Hersey sponsored four exercises immediately following his appointment at the Program Manager of LRAPP. One exercise, the Pacific Acoustic Research-Kaneohe/Alaska (PARKA), collected some excellent data. In fact, PARKA may be the best oceanographic data set ever taken, even to this day. The other three exercises were measurement disasters. Generally, disasters mean the death of the program. But, in this case, these disasters transformed the LRAPP administration and focus to one that eventually led to success in every part of the program.

PARKA

Literally hundreds of people were associated with the PARKA series of experiments. There were three experiments as part of the PARKA series. The care and devotion of the participants in gathering the data associates with this major oceanographic measurement series set a standard for at-sea experiments that is still followed today. The acoustic database for the PARKA series of experiments remains unique among oceanographic databases because of its exacting detail and accuracy.

The first major measurement experiment took place in the Northeast Pacific. PARKA was run of over 1700 nautical miles from Hawaii to the Aleutians to validate the acoustic models. The Chief Scientist was Dr. Rudy Nichols from AT&T.

Seven units were scattered throughout the operating area to gather acoustic data. The only way to integrate the measurement results was through rigid adherence to a time schedule. After the experiment and during the ensuing months, the data reduction and analysis could be coordinated only by knowing the location of each unit at the specific time. Throughout the experiment, many of the units at sea never came close to each other.

PARKA Timetable

PARKA I	August–September 1968
PARKA IIA	September–December 1969
PARKA IIB	March 1970

The PARKA program had three components: PARKA I, IIA, and IIB were all run roughly to the north and northeast. All the measurements were taken along lines of bearing from Hawaii to the continental shelf of North America.

Environmental measurements were taken along the dsignated line of bearing and consisted of all measurements relevant to underwater acoustics. The standard oceanographic measurements were temperature and salinity as a function of depth. These data were supplemented by measurement of the speed of sound (sound speed) directly, also as a function of depth. Further, the bottom profile, i.e., the bathymetry, was measured in detail.

Acoustic measurements consisted of placing hydrophones in a vertical string and measuring the amplitude of the arriving signals. The sources of the signals were explosive charges dropped from a ship (later airplanes). These explosives are called Sound Underwater Signal (SUS) charges. The drop time of the SUS charges was measured and noted with great precision. The arrival times for a given range had been precalculated, so that with a reasonable degree of certainty, the arrival time for the SUS-generated acoustic wave could be correlated with when and where it was dropped.

As a simple example of why this information is so important, consider the travel time for an acoustic signal. Sound travels at approximately 5500 feet per second in water. Since 1 nautical mile is 6000 feet, it takes an acoustic signal approximately 1 second to travel 1 mile. When a SUS charge is detonated at a range of 300 miles, the signal takes approximately 300 seconds (or 5 minutes) to arrive at the receiving hydrophone.

To complicate matters, sound does not travel in straight lines; it follows curved paths that are dictated by the speed of sound in the ocean at each point along the propagation path. Because the paths are not straight and the speed of sound varies with depth (as well as with range), the arrival of the SUS signal is not a single precise arrival. Rather, the arrival of the acoustic signal sounds like a rumble, similar to thunder. This sound indicates that while all the signals started at the same time, different paths and thus different parts of the acoustic signal arrive at different times. Therefore, great care must be taken to associate the arriving spread-out acoustic signals with a particular SUS event. This task was not easy, but with tremendous care on the part of all participants, it was done.

The knowledge of the signal strength of the arriving acoustic signal and the knowledge of the signal strength at the SUS explosion (or actually one meter away from the explosion point) allows the transmission loss to be measured. If you use a sufficiently accurate acoustic model, the transmission loss can be correctly calculated. Comparing the measured

transmission loss and the calculated transmission loss allows an evaluation of the accuracy of the computer model. Further, since the computer mod-el requires environmental inputs (both the sound speed field and the bathymetry must be known), this process of model evaluation is difficult, time consuming, and dependent upon a myriad of details.

Two additional measurements must be made: ambient noise and detailed information about the operation of the measurement system itself. During PARKA, since all the hydrophones were recorded, the ambient noise[29] was recorded as well. The ambient noise measured during thePARKA experiment was not considered of great importance, since the development of ambient noise models was still in the conceptualization stage. There were no ambient noise models which could predict ambient noise, and therefore, there were no mathematical predictions available to compare to the measured data. The computing requirements for making ambient noise calculations were simply too great for the state of the computing equipment at that time.

The measurement systems were simply omnidirectional hydrophones. While there were many of them, they were not considered as components of a distributed array. However, due to the detailed knowledge of the hydrophone distribution, and their individual performance, it would be possible to make virtual arrays mathematically, and make predictions on how such arrays, if they existed, would perform.

---

[29] The acoustic signal measured while there was no SUS shot data arriving.

How I Got to Make Frequent Trips to Denver

Denver has mountains. For a Texas boy, mountains are a big deal. Fortunately for me, Texas Instruments knew very little about direct record technology. Actually, none of us knew about direct record technology. Woods Hole had chosen the direct format because it represented the highest information density available. They also had chosen a seismic IRIG tape recorder developed by a contractor that was capable of continuous recording for periods of up to 10 days. This tape recorder was not the best on the planet. It recorded at a speed of 15/160 inches per second, which is so slow you can hardly see it move. However, it could record for 10 days; it was 10 years before we found a replacement. It turned out that the direct record format required a lot of care and attention to be successful.

For a time we received tape recordings of variable quality, and it was clear something had to be done. The program office somehow came up with two expert engineers at Martin Marietta: Leighton Meeks and Glen Schulze. I believe Roger Harris from NOSC was instrumental in finding them. They did two things. They trained the Texas Instruments engineers on how to set up the master tape recorders, and they also provided high-quality copies from the Denver facility.

My job was to pick up the master tapes when the ships would return to port and deliver them to Denver. We would make strip charts while making the copies, so we could immediately give a quick report on data quality. It was great.

Jack Shooter

SEA SPIDER

SEA SPIDER, attempted in 1969, was designed to be a horizontal suspended array that would be in the sound channel. It was clear to Dr. Hersey and the underwater acoustic community that the acoustic signals, even from relatively close sources, interacted with the bottom and became distorted. Since the acoustic signals propagated very long distances (thousands of miles) and was undistorted when the energy remained in the sound channel, it was obvious that putting a measurement or detection system in the sound channel was a good idea.

Such a system had never been attempted, but it was a measure of the desperation of the U.S. Navy to solve the ASW problem that the experiment was given support. The SEA SPIDER array, however, was so complicated that it couldn't be deployed. The experiment was to be in the Northeast Pacific, because PARKA data had shown that this area was good for collecting data. In addition, Soviet submarines were operating in the Pacific. Unfortunately, due to the complexity of the system and the inability to deploy it, no data were obtained.

At that time, and to some extent today, if a major measurement operation is attempted and it fails, the program is canceled. This process seems harsh, but funds are always limited and other programs are always competing for those funds. If a program is considered to be of potential value to the Navy and has vocal and persuasive champions, then it generally has one chance to demonstrate its value. It is significant that even with the complete failure of SEA SPIDER, another experiment was attempted the following year in the Atlantic.

TEST BED

The next effort, known as TEST BED, took place in December 1970. Instead of making the measurements with a simpler system and recording the data at sea, it was decided to use a portion of the SEA SPIDER sensing equipment, and then cable the TEST BED system data back to Bermuda. The recording equipment at the termination of the underwater cable was located in Bermuda.

Unfortunately, this experiment was also a failure. Data were received for only 15 seconds. The LRAPP program faced cancellation. While the U.S. Navy was becoming more desperate and VADM Martell was still a strong supporter, the program had produced no results of any value. Its existence was in jeopardy.

Other Developments

At ONR, Dr. Marvin Lasky had directed the development of the towed array, which is a relatively large number of hydrophones placed in a line and towed behind a ship. The hydrophone spacing was dependent upon the frequency being investigated. The concept of staggering the spacing and having different geom-etries so that several different frequencies could be considered simultaneously was still in the future.

Dr. Lasky, supported by Dr. Folger Whicker, decided that this system would work. It would allow a single line of hydrophones to have their received signals mathematically added together, which would increase the probability of detecting an acoustic signal. Further, if the signals were added together but staggered in time, they could "look" in different directions. This concept was new and quite startling. True, such efforts were being done in the radar world, but this was acoustics and therefore different. These towed arrays, called the Interim Towed Array Surveillance System (ITASS), were essentially movable SOSUS systems that would be pulled by U.S. Navy ships.

The Navy surface ship community, however, simply wouldn't hear of it. The ITASS arrays had to be pulled at only a few knots, or else the noise generated at each hydrophone, due to the flow noise, would just be too high. Destroyer captains were taught that their destroyers were multi-mission ships and were supposed to operate at high speeds, e.g., 25 knots—only now these ONR scientists were suggesting that they throttle down to less than 5 knots and simply pull a long hose behind the ship. The submarine community already had some small towed arrays and found that the arrays were helpful. Navy officers learn to follow orders, even if they find those orders repugnant. The three destroyer escorts, the USS *Lester*, the USS *Courtney,* and the USS *Hammerberg*, reluctantly towed the ITASS arrays at 5 knots or less.

In addition to these developments in ONR, the Defense Advance Research Project Agency (DARPA) had begun the development of a very special and unusual array. This towed array was called the Large Acoustic Marine Basic Data Array (LAMBDA). Dr. Henry Aurand had conceived the idea of LAMBDA. It was to be a towed array 15,000 feet (3 miles) long. This array could resolve directionality of signals, as well as detect very low frequencies.

Mobilization: LRAPP Style

The M/V *Indian Seal* was the support ship for the Large Acoustic Marine Basic Data Array (LAMBDA) array. She was a 210-foot LOA supply boat: bridge and quarters forward, the remaining aft 130 feet was clear aft deck. To convert the *Indian Seal* into the LAMBDA support ship required installing 10 ISI 8- × 8- × 40-foot containers and two ISO 8- × 8- × 20-foot containers, two winches, including the 120-ton LAMBDA winch with its overboard sheave, deck conveyor system and level wind, Grove crane, a 24-inch raised deck, and miscellaneous subsystems. Five days were scheduled for basic mobilization, then four days of testing various systems and preliminary sea trials, then back for five more days of final mobilization, including plumbing, electrical, special communications systems, and others. Anyone with mobilization experience can appreciate that the LRAPP Mob schedule was ambitious, but we met it.

I had won the contract to mobilize the *Indian Seal*. I leased dock space in Galveston, Texas, and brought in all the necessary trades. To load the ISO containers and other equipment, I leased a 150-ton mobile crane. We were able to install all the equipment on board except the LAMBDA winch. We had a problem there.

The winch came from NOSC San Diego and was trucked to Texas on a double-drop, extra-wide trailer. When the winch arrived, it looked good, clean, and newly painted. During the installation process, when the winch was 20 feet in the air, I noticed that the winch flanges did not look as they should. Apparently, during the transit, the winch must have hit an overpass. It was badly out of line, so much that the drum could not rotate. A new winch drum was out of the question because of schedules. Discussions with my welding crew resulted in a possible solution. We welded pad eyes on the flanges, installed chain falls, and positioned 12 welders with "rosebud" tips heating torches to heat the 1.25-inch-thick steel flanges and the ribbing. We heated and selectively pulled in on the chain falls. After 18 hours, we were able to restore the winch to operating condition. The cost of the repair was $5,000 for an extra day with the mobile crane. The mobilization schedule was not compromised.

I mobilized the LAMBDA in March. In November, the array was lost off Bermuda. In December, I demobilized the M/V *Indian Seal*.

Claude P. Brancart

Chapter Seven
LRAPP-Restructuring

General Comments

The failure of the third at-sea experiment caused Dr. Hersey to make changes in personnel and the way the program was run. He hired Dr. Roy David Gaul as the new Program Manager. Dr. Gaul had a completely different management style from Dr. Hersey. He was extremely well organized and believed that the program should be run with tight management controls.

Dr. Gaul was an extremely competent scientist and engineer. He clearly understood that much of what was being attempted was high risk and thus subject to failure. He never held team members responsible for failure of physics, but he held the entire team accountable for performing at their peak capability and for keeping him and his staff informed about progress and difficulties at all times.

> **Management Styles**
>
> The qualities of management demonstrated by the example in the following EASTLANT section are not completely typical of Roy's management style, but they are not rare either. A colleague answered the question, "How would you characterize Dr. Gaul's management style?" with, "You might call it the Kung Fu style of management."

To some extent, Dr. Gaul could be compared to a symphony orchestra conductor: he led the orchestra and integrated the con-tributions from each member of the orchestra. Similarly, he specified and integrated the relative contributions of each team member. He demanded excellence from everyone and, because of his unwavering support and loyalty to his staff, he received the best that each team member could provide. The result of his leadership was a series of unprecedented successes for the U.S. Navy and the oceanographic community.

Each contract, whether with an industrial organization, a Navy laboratory, or a university, had a precise and detailed Statement of Work

(SOW). As part of this SOW, the deliverables were specified in great detail. In addition, a carefully thought-out schedule was associated with the performance of SOW. The entire mood of the program was changed from one of academic interest and scientific inquiry to one of delivery of fundamental ad-vanced development systems that would work, and work on time.

Finally, Dr. Gaul insisted on a firmly interlocked team approach to the entire prob-lem. Personality conflicts were simply not allowed. Everyone did their job, on time, and with satisfactory results. There was neither room nor time for lack of coordination. Approximately 10% of the team participants were dismissed every year for the first several years until it became clear to all that they must work together.

Roy Gaul—Opening Shot

Prior to reporting in January 1971 as the manager of LRAPP, the project had enjoyed a free flow of funds in spite of some significant equipment failures. A so-called TEST BED project was ongoing as the latest attempt to install a suspended low-frequency array in the oceanic sound channel. It was installed in November 1970, but the recording station at Tudor Hill in Bermuda was not receiv-ing acoustic signals. In December OP-095 rejected a LRAPP plea for supplemental funding in the amount of $1.8 million to continue efforts to bring the system on line. Dr. Brackett Hersey issued a stop order in December to the Supervisor of Salvage and the Naval Undersea Sound Center (New London, Conn.), which were key partici-pants. When I reported for duty, there was no evidence that the order was taken seriously; work was continuing, presumably with the expectation that, as before, LRAPP would find money. Unfortunately, the LRAPP funds had been committed early in the fiscal year, and the recipients were loathe to give refunds. There was no flexibility for reprogramming.

Roy Gaul—Opening Shot (continued)

By mid-January, I had assembled a cable diagnostics specialist (courtesy CAPT Joe Kelley, SOSUS program manager) and some technical advisors for a trip to Bermuda. It took two or three days to determine that there were several discontinuities in the vicinity of the array. At the end of the week, we had a review meeting before returning to D.C. The technical team emphasized the singular accomplishment of installing the system in rough seas with the planned geometry and without casualties. I asked if they considered the project a success. The response from the team leader was yes. My rejoinder was that it was viewed as a failure by LRAPP because the system produced no electrical wigglies. They were shocked to learn that the funding was depleted and the project terminated. I offered to meet with their managers at NUSC in New London the following Monday.

The meeting was not cordial. The heretofore unknown new manager of LRAPP held his ground. The Supervisor of Salvage, with some rancor, had to absorb a cost overrun of about $700,000. To the best of my knowledge, the TEST BED system remained silently intact throughout my tenure and may still be a suspended relic of the often messy process of technology development.

Roy Gaul

Navy guidance was highly interactive. LRAPP was to solve the ASW problem using the scientific community. No organization in the country had closer contacts with the scientific community than ONR. At that time, LRAPP received administrative guid-ance almost daily, as well as restructured and defended itself two to three times a year.

LRAPP decided to abandon the quasi-operational system construction and con-centrate on models, data-bases, and measurements. It was the LRAPP concept that proper use of the environment could result in a 10–20 dB[30] increase in the S/N, while equipment gains generally were only 2–3 dB..

In the 1970s, there was substantial debate in the U.S. Navy about the impact of surface ship noise. While many people thought that the ship noise filled in the notch between blade rate (the frequency related to the whirling propeller, and, of course, is related to the engine speed) and approximately 100 Hz, it was also believed that the noise was omnidirectional due to the number of merchant ships at sea. It was felt that the ambient noise is that frequency band would be continuous, but there were specific frequencies (lines) that were clearly present and due to surface ships. Further, it was believed that the noise was stationary and white. Neither assumption is true. Low-frequency noise is dynamic, and it is not white. The details of the characteristics of this noise became very important in 1971 and 1972.

Some scientists at Bell Laboratories[31] believed that the noise was omnidirectional. It was believed that regardless of which direction the measurement was made, the value for the noise would always be the same. The measurements of ambient noise for the eastern Atlantic were very high. If the noise was omnidirectional, then the effect would be that the S/N would be so low that submarine detections would be made only at short ranges. AT&T scientists advised against implanting the SOSUS array for the eastern Atlantic.

But, if the noise was directional, and a much lower noise level could be sensed if the senor could look in a particular direction, then the S/N would be much higher and detections of submarines could be made at long ranges. LRAPP insisted that the noise was directional.

The noise prediction models worked quite well for frequencies at 1000 Hz and above. But, the noise prediction models for low frequencies, about

---

[30] The term dB, an abbreviation for the measurement of decibels, is a mathematical definition. Specifically, it is a logarithm: dB=10 $\log_{10}$ (number/base). For example, if the ratio (number/base) =2, then this value calculates to 3 dB. Thus, if one number is 3 dB larger than another, it is twice as large. If a number is 10 dB larger than another, it is 10 times as large. A 10 dB increase in signal detection is enormous. The term Bel is named in tribute to Alexander Graham Bell. A dB, or decibel, is one-tenth of a Bel.

[31] Bell Laboratories designed the SOSUS. They also advised the Navy on how to improve performance and tried to explain why things didn't work. Western Electric was on call to repair broken cables and fix other things that failed.

50 Hz, were not good. Since the detection of submarines was made at about 50 Hz, the value of the S/N was critical to know and predict. Operationally, the implications were serious. At that time, there was no way to find Soviet submarines from the Greenland– Iceland–United Kingdom (GIUK) gap to the Straits of Gibraltar prior to entering the Mediterranean Sea. If the submarines could not be detected, then Soviet submarines could appear suddenly in the Mediterranean, and possibly putting the United States naval forces at a disadvantage, and possibly at great risk. The problem was critical, and a decision had to be made about implantation of SOSUS arrays in the Atlantic.

The Navy had to decide whether or not to implant the array; since they could not be sure that the array would help solve the submarine detection problem, and they did not want to spend an enormous amount of money on a system that would not work, they hesitated. They decided to postpone the implantment for a year.

Dr. Gaul convinced VADM Hal Shear[32] that LRAPP could take directional ambient noise measurements using the ITASS arrays. These arrays were being operated in the Mediterranean by the USS *Hammerberg* and the USS *Courtney*. Dr. Gaul insisted that the SOSUS could be a directional system if they incorporated beamforming, and if the noise was directional. After some hesitation, VADM Shear agreed that the experiment would take place in the eastern Atlantic. It was to be called the EASTLANT exercise.

If AT&T scientists were correct, then the noise measurements would show that the ambient noise was omnidirectional—a finding that would essentially doom the use of the SOSUS arrays for long-range detection of Soviet submarines. It would also mean that Soviet submarines could operate virtually undetected over most of the world's oceans without fear of being attacked. Many people believed that the SOSUS system, which was very expensive to construct and deploy, would simply not give a reasonable return on investment. The entire surveillance system was in jeopardy.

If Dr. Gaul and LRAPP scientists were correct, then the noise measurements would show that the ambient noise was directional. This finding would mean that the installed SOSUS arrays could be used in an entirely different way than they had in the past. Further, if the noise was substantially directional, then the Soviet submarines could be conceptually detected anywhere in the world's oceans. *In few instances had the scientific*

---

[32] The successor to VADM Martell as OP-095.

community been involved in making a measurement and providing scientific information that could spell the difference in how the Cold War would proceed.

EASTLANT

VADM Shear and Captain Vern Anderson, U.S. Navy, who was PME-124[33] at that time, agreed to provide the towed arrays and their surface ships which were necessary to measure the directionality of the noise.

Raisbeck[34] and Williams invented the measurement technique to ascertain the directionality of the noise field using two towed arrays. The towed arrays available were the ITASS arrays. They would be pulled at very low speeds (3–4 knots) in very specific patterns. The patterns were octagonal (a figure with eight equal sides), and each side was long. The use of this type of pattern approximated a circle. Additional patterns with 12 equal sides were also used.

One leg took approximately 1 to 1-1/2 hours to complete. About 3–4 days was spent at each site. Only the amplitude of the noise was to be measured, not the phase. This measurement procedure would provide sufficient data to answer the noise directionality question. The experiment, which was to last for two weeks, would be under the control of Dr. Gaul, a civilian, something that was completely outside the experience of the operating Navy at that time.

The position of the Sixth Fleet was adamant: they would not give up use of the operational towed arrays in the Mediterranean to operate under scientists in the eastern Atlantic. It was the power of OP-095 and PME 124 that forced the issue. Not only were the ambient noise measurements made during the EASTLANT exercise, but environmental measurements that were necessary for the inputs to the models were also made. Shipping surveillance was made to learn the location of most of the merchant fleet operating in the eastern Atlantic.

EASTLANT was a two-week experiment. The results showed an approximately 10-dB azimuthal variation in the directional noise, which could be directly correlated to the nonuniform, but relatively static, shipping

---

[33] PME is the acronym for Program Manager, NAVELEX; the organization in NAVELEX was 124.

[34] Interview with Gordon Raisbeck and Budd Williams.

distribution. The results were presented to VADM Shear. After considerable discussion, it was decided to permanently implant the SOSUS system in the eastern Atlantic at a cost of approximately $180 million. AT&T made the necessary equipment changes in order to allow enhanced analysis of the signals. The philosophy of the SOSUS system was changed. The SOSUS system went from detecting signals in an assumed omnidirectional ambient noise field to detecting signals by taking into account the directionality of the measured ambient noise. This effort was quite substantial. The changes were made based upon predictions from the models and measurements made in EASTLANT. It was a significant accomplishment that the SOSUS system worked just as predicted. For the first time, Soviet submarines could be detected and continuously tracked from the GIUK Gap to the Straits of Gibraltar. This change to the SOSUS system and the associated analysis changed the entire operational procedure used by the United States Navy in dealing with the Soviet submarine force. It made the Soviet submarine force vulnerable to attack.

As a direct result of this measurement exercise, the credibility of LRAPP was established with the senior officers of the U.S. Navy who were responsible for ASW. The LRAPP scientists had shown that they could provide something of value to the operating Navy. LRAPP was virtually assured of continuing support from both OPNAV and senior Navy civilian leaders. The substantial enhancement of the SOSUS system through the forceful contribution of LRAPP was a notable achievement. But, in itself, it did not solve the ASW problem. There was much more to accomplish, and LRAPP applied itself to the detailed problems with vigor and determination.

Chapter Eight
LRAPP-Continuing

General Comments

LRAPP enjoyed spectacular success with the EASTLANT experiment. Dr. Gaul and the members of the LRAPP leadership had faced the predisposition of the existing power structure and demonstrated that ASW problem might be solvable if the environment was used correctly. Senior Navy officers within OPNAV now agreed that the scientists might be of use to them. The Navy civilian leadership was delighted with the success of the program and the suddenly enhanced value of a costly investment, the SOSUS system. Now it remained important to make the rest of the U.S. Navy associated with the ASW world believe that the scientific community could contribute something to the operational world.

An Unusual Computing Problem

During TASSRAP, several other people provided support in the basement of the Marine Barracks in Naples, Italy. For example, the TASSRAP model was run on a NOVA 800 computer. It was designed to operate at 60 Hz using 110 volts. In Italy, how-ever, the power is 50 Hz and 220 volts. Thus, the power requirements for the NOVA 800 computer had to be met by running an external generator outside the Marine Barracks and snaking the power cables through an open window. Today, any mediocre laptop is far superior to the NOVA 800 and is designed to be powered by any voltage and frequency that exists as standard anywhere in the world.

TASSRAP, DECKPLATE, and the MED ASW Augmentation Program

In 1972, the three ITASS towed arrays were working in the Mediterranean Sea. TASSRAP and DECKPLATE were scheduled as LRAPP exercises in the Mediterranean area. These exercises were different in character from each other. TASSRAP was principally a scientific exercise.

Its purpose was to gather scientific information in support of the LRAPP, and ultimately to support fleet ASW operations. DECKPLATE was a much more focused fleet exercise, in which the three ITASS arrays would be used by fleet personnel, with some slight coaching by LRAPP personnel (or people who were under contract to LRAPP) to provide support.[35]

The required scientific measurements were the transmission loss and ambient noise in the basins where the exercise took place. The principal focus of TASSRAP was to obtain the scientific information necessary to help evaluate and validate the TASSRAP computer model. This task was the responsibility of Mr. Jim Gottwald. He worked jointly with Mr. George O'Sullivan, who was responsible for the fleet acoustics by ensuring the accuracy of the reported contacts. TASSRAP required the services of U.S. fleet submarines to perform certain measurements. This meant that the submarines had to be in certain places at certain times, etc. The coordination of all the units which supported TASSRAP was substantial. Each operational unit, naval and civilian, had to frequently report their position. It was Mr. O'Sullivan's job to make sure that all of the reports were consistent. After the entire exercise was complete, it was necessary to reconstruct what each platform was doing at a particular time. This was a very complex job. It was crucial that the spatial geometry of the entire exercise had to be correct if the measured data, taken by the individual units, was to be understood. The re-construction of all the platforms, their positions, operational evolutions, and other performance characteristics was the responsibility of Mr. Sam Elam. These three men worked as a team.

Three ships towed the ITASS arrays: the USS *Courtney*, the USS *Hammerberg*, and the USS *Lester*. Their required speed of advance was less than 5 knots, and they were not happy about their role in the experiment because they considered themselves to be multi-mission combatants and, indeed, they were. Towing pieces of hose at 5 knots was not the role for them to play but, being dutiful Navy ships, they did what they were ordered to do. The information was gathered and a detailed report written by the LRAPP men on site,[36] in itself, an unusual procedure.

---

[35] Interview with Mr. Sam Elam, Mr. George O'Sullivan, and Mr. Jim Gottwald. This interview was conducted by CAPT Terry J. McCloskey, USN (Ret.), who was deeply involved in support of LRAPP fleet operations as a member of the LRAPP staff.

[36] It was truly LRAPP "men" at that time. Not until later in the program did women become contributing members of the LRAPP team.

Generally, exercises were conducted and then the weary participants went home to rest, relax, and contemplate their adventures. Only much later did they write a report. Dr. Gaul ordained, correctly, that no one could come home until the fundamental complete draft report was written and reviewed by the Navy hosts. This mandate caused the report to be written as the exercise proceeded, resulting in much more detail, more accuracy, and with a much more timely deliverable than the operational Navy was used to or expected from the scientific community.

The frequency used for making detections at that time in the Mediterranean was 50 Hz. The real trouble, however, was that the ambient noise at that frequency was so high that detections were difficult to make, particularly at long distances. In the Mediterranean, long distances are considered to be approximately 100 nautical miles. The three destroyer escorts did their jobs well as shown by the reports. The destroyer captains grudgingly admitted that maybe the scientists did know something, but a multi-mission capable destroyer was still not the proper platform to pull towed arrays. These two exercises, TASSRAP and DECKPLATE, were the beginning of successful modern ASW operations in the Mediterranean Sea.

The day for the fleet was at hand. With the success of EASTLANT, which occurred in 1972, as well as TASSRAP and DECKPLATE, VADM Shear and CAPT Anderson decided to push their luck. After EASTLANT, the three destroyers and their ITASS arrays had remained in the Mediterranean. But, their success in detecting Soviet submarines was poor. Finally, after two years, VADM Shear decided to pull the ITASS arrays out of the Mediterranean, and decided on another approach to find Soviet submarines.

In 1973, Dr. Gaul, Dr. Hersey, and then-Lieutenant Commander (LCDR) Terry J. McCloskey were called in for a high-level meeting. This was going to be a discussion about a very clear mission. While Dr. Gaul and Dr. Hersey were used to high-level meetings, it was certainly unusual for LCDR McCloskey to be invited. He was not exactly sure why he was accompanying his superiors to a meeting with VADM Shear; he was to learn shortly.

While it is amusing to relate now, over thirty years later, LCDR McCloskey had his troubles at that time. It is fine for a vice admiral to tell a lieutenant commander to do something; it is not so easy for the lieutenant commander to make things happen. LCDR McCloskey had two months ro get the ships ready; the transit from Charleston to the Med-iterranean was to take one month. However, the LRAPP team worked together, and in January 1973 the program began in earnest.

LCDR McCloskey was tasked to design a major ASW program for the Mediterranean Sea and get back to VADM Shear with the plan within 24 hours. This short time frame was (and is) unprecedented in designing such a complex mission. LCDR McCloskey imme-diately and frantically contacted the appropriate LRAPP team members, and the first meeting was held in the pre-dawn hours in the parking lot outside of Building 58. This was a serious, important program; it was going be treated that way. No efforts were spared, including meeting in a parking lot in the pre-dawn hours. .

Five tasks were decided upon almost immediately. These tasks were a strange melange of science, academics, and naval operations.

- *Review all Mediterranean environmental information*

The first task was to review all the information that existed throughout the Navy, with particular emphasis on FNWC, and such academic institutions as Woods Hole Oceanographic Institution, Scripps Institute of Oceanography, and others. This effort resulted in a classified atlas of Mediterranean environmental information. The resulting atlas clearly showed what we knew, what we had doubts about, and what we didn't know.

- *Employ the TASSRAP model to provide performance predictions for all the existing fleet systems throughout all the ASW prediction areas.*[37]

The second task was straightforward in the sense that the TASSRAP model was exercised. TASSRAP made performance predictions for every system used for ASW purposes by the operational navy, and every system used by LRAPP for each ASW prediction area. However,

---

[37] ASW Prediction Areas are described Chapter 13 in the section on Model Development and Operations.

there were many ASW prediction areas located in the Mediterranean, and many different systems were used by the operating fleet, and by LRAPP. The number of calculations were enormous; further, a system had to be designed to allow easy storage and access of the calculation results by the users of the predictions.

- *Deploy bottom systems in the Straits of Sicily.*

The third task was to investigate the eastern side of the Straits of Sicily to ascertain if there was an acoustic advantage for VP and/or towed arrays. The Acoustic Data Capsule (ACODAC) system was deployed and the results were very encouraging. This effort was followed up by a later exercise, SEINE NET, which developed operational tactics necessary to exploit this acoustic advantage.

- *Revitalize the two minesweepers, USS* Alacrity *and USS* Assurance.

The fourth task was to reactivate two MSOs: the USS *Alacrity* and the USS *Assurance*. These minesweepers were normally employed as single-mission vessels and were design to proceed very slowly for long periods of time. Although they were ideal platforms for the towed arrays, they certainly were not built for such a task as providing support for the towed arrays and all their ancillary equipment. Furthermore, their captains, young lieutenants, did not rank highly in the U.S. Navy command hierarchy. They would be required to perform some serious tasks and were not accustomed to being in the spotlight.

- *Employ the LAMBDA array for scientific purposes, obtain data in the Levantine and Ionian basins, and ascertain directional noise.*

The fifth task was to use the LAMBDA array for scientific purposes. This task would require TASSRAP to make predictions for placing the array and its platform in the right positions while looking in the optimum directions to enhance the probability of detecting Soviet submarines. The placement of the LAMBDA array was highly dependent upon the Mediterranean environment. The direction it was towed would depend upon the characteristics of the directional ambient noise.

These five tasks were decided on before the morning was fairly begun. LCDR McCloskey was then sent to see VADM Shear and CAPT Anderson. To his amazement, all of his suggestions were accepted without change. He was ordered to implement his plan.

There were, however, some problems. The minesweepers were tied up, unused, and awaiting mothballing. For towed array use, two ships were needed to allow for cross-fixes to be established. Further, they had to proceed slow and steady, which is the normal mode of operation for minesweepers. The two towed arrays were refurbished as Advanced ITASS systems.

The members of the exercise consisted of both military personnel and environmental scientists. The exercise could be successful only if joint training took place. That training was given to the minesweeper crews and the VP[38] crews in Norfolk, Virginia. Everyone involved learned how to do cross-fixing. Two people in each ship were assigned to tell the crews where to place the arrays. In addition, all the crews needed to know the Mediterranean environment. ASW in the Mediterranean had been a failure; this exercise began a new ball game.

From January to August 1973, the ships operated during six cruises in the Mediterranean. TASSRAP was the model for this exercise. Dr. Ed Tunstall and Dr. Ron Wagstaff were the principal investigators on board the ships. Since these ships were not assigned to the Sixth Fleet, they had no preplanned logistics support. Everything that was needed was furnished by a submarine tender in Rota, Spain.

The ships searched the Levantine Basin. The detection ranges were about 200 nautical miles. At first, there were no contacts—not an auspicious beginning. But fortune smiled upon the effort, and they began to detect submarines. Not only did they begin to detect submarines, they also obtained a complete set of tones. At that time, the U.S. Navy was accustomed to classifying on one or two tones, but previous doctrine had required three.

Cross fixing was done by LT Steve Jones in the *USS Lester* during the DECKPLATE exercise in 1972.This skill was brought to the Mediterranean ASW Augmentation Program in 1974. Also, during 1972, it was noticed that Soviet submarines used sunken wrecks for navigation points. They frequently came to a wreck before starting out on a new course. In addition, three surfaced Soviet submarines were visually confirmed.

The Soviet submarines left the basin and traveled to an Egyptian port. The Soviets apparently remained unaware of the U.S. Navy's presence in the area. There was always the issue of covertness, but the Soviets never

---

[38] VP is an abbreviation for multi-engine ASW Maritime Patrol Aircraft. They operated out of many places in the Mediterranean, including Rota, Spain, Sigonella, Sicily, and Naples, Italy.

seemed to discover that there was something special about these two minesweepers.

From these cruises, six fleet reports were issued by Commander, Task Force 67. For such reports to be published by the command was highly unusual.

Aftermath

This exercise changed the paradigm on how detection and classification were made; in addition the exercise established credibility with the fleet. Previously, detection was attempted only at 50 Hz, but in the Mediterranean they began looking for 300 Hz lines. The ambient noise at 300 Hz is substantially less than noise at 50 Hz. Therefore, for a given signal strength, the S/N is higher for 300 Hz than it is for 50 Hz.

Control of the assets was transferred from the Atlantic Command to the Sixth Fleet. Both the USS *Alacrity* and the USS *Assurance* had gained respect by their successes. The two ships, in three months, had over 671 hours of verifiable contact time against Soviet submarines.

The Sixth Fleet was now desperate. The original ITASS arrays, towed by the USS *Courtney* and USS *Hammerberg* had failed in the sense that they simply did not make or maintain contact with Soviet submarines, *but the minesweepers worked*. Now, they had to return to the United States, leaving no reliable asset in the Mediterranean theater.

In the end, the *Alacrity* and the *Assurance* returned to United States home waters and were decommissioned and sold for scrap. The cost for refurbishing them as minesweepers was too great. But the fleet operators had seen what could be done. Also, someone else recognized that the game had been substantially altered—the Soviets realized that they were in danger of being detected in the Mediterranean Sea. They left the area for a substantial period of time.

After the success of TASSRAP, the LRAPP charter was upgraded to include fleet support. LRAPP received additional funds from OP-095 to continue these types of support and demonstration exercises.

In the meantime, the exercise of the LAMBDA array provided LRAPP with substantial environmental data. LAMBDA was a new towed array designed for the Advanced Research Projects Agency (ARPA), which eventually became the Defense Advanced Research Projects Agency (DARPA). This array, the brainchild of Dr. Henry Aurand, was much longer than earlier towed arrays. The length allowed it to be sensitive to very low frequencies, as well as generate very narrow beams. This array was now being used in the Mediterranean for scientific measurements. An offshoot of LAMBDA was to arrive in the Mediterranean several years later and achieve astounding success.

CHAPTER Nine
Meanwhile...

General Comments

While LRAPP was building equipment, going to sea, making measurements, and reducing and analyzing data, another part of the Navy closely aligned with LRAPP was also undergoing substantial growth and success: FNWC. The modeling community, in support of LRAPP, and indirectly throughout the USN, began to deal directly with FNWC.

More About FNWC[26]

FNWC's success depended on many things, but most certainly on the existence of mammoth computing power. The development of the digital computer and its effects has been told by many. In this particular instance, it was the advent of the IBM 704 that affected FNWC.

In 1954, the Naval Weather Service began to make daily weather predictions using the IBM 704. In 1956, NAVSEA issued a contract to a new company called Control Data Corporation (CDC), which was begun and developed by three former Navy officers, Dr. William Norris, Dr. Seymour Cray, and Dr. James Thornton. Their contract with ONR was to design and build the first all-transistor computer (later known as the 1604).

The first two computers were built and installed in the new FNWC facility in 1958. At that time, the entire staff consisted of CAPT Wolff, two ensigns, three civilians, and six enlisted personnel.

FNWC was the only place in the U.S. Navy where world-wide ASW predictions were made on a 24-hour-a-day basis. CAPT Wolff points out the reasons for their uniqueness:

- FNWC had the computers to do what was required.

---

[26] Interview with Captain Paul Wolff, USN (retired).

• FNWC had databases for the Northern Hemisphere included water temperature and salinity as a function of depth, the water depth in a 6-mile grid, and the bottom reflectivity for low-frequency sound.

• FNWC had the organization, the communications equipment, and the motivation to support the fleet on a timely operational basis.

Everything occurred because of a confluence of people and events, most of which were accidental and unique.[27]

In 1967, FNWC was directed by CAPT Wolff, a world figure in meteorology. Two CDC 6400 computers were located at FNWC as well as CDC 1604, Serial No. 1. This machine used a 10-character word. It was the fastest machine in the world at that time. While IBM had led the digital computing world for many years, CDC was specifically designing its machines for scientific, rather than general, work. The CDC 6400 was an excellent resource. It was used for the development and refinement of weather calculations. Remember, FNWC was an operational command: computer predictions were made and transmitted to the Fleet on a regular basis.[28]

Working Environment

The working environment was remarkable. All members of the LRAPP community and others associated with FNWC worked as members of a team. If somebody decided not to cooperate, then Dr. Gaul made sure that they were no longer part of the workforce.

[27] Ibid.

[28] Interview with Mr. Ken Osborne

FNWC's Early Computers

FNWC was a two-story structure similar to a block house. The entire building was devoted to supporting the CDC computer, which ran on a 24-hour-a-day schedule. There was no reason to distinguish day from night and, thus, the building had no windows.

By the late 1960s, FNWC had two CDC 3200s, one CDC 1604 Ser 1, and two CDC 6400s. One entire wall had an array of tape units. Texas Instruments, which was supporting FNWC, had an IBM 360. This compu-tational effort supported fleet predictions. It required the computer be gently closed down nightly, repro-grammed, and started up as soon as possible. This operation took all night, but the computer was completely reprogrammed and back on line before the morning calculations began.

When FNWC received the CDC 6400, it was a world-class computer. All programs were written in machine language using fixed-point arithmetic.

Due to its superb support of the operating fleet, FNWC was awarded the Outstanding Contributor Award by the CNO on 10 December 1962.

At that time, the ASW emphasis was on digital signal processing (DSP). To make advances in DSP, computer time was extremely important. Computers were not readily available then, nor were they easily accessible for experimentation and development. Two fleet models had been developed. One was the Acoustic Range Prediction Systems (ASRAP) model for VP developed by CDR Tatro, The other was the Ship Helicopter Acoustic Range Prediction System (SHARPS) model for ships and helicopters developed by LT McCloskey. Predictions were provided to the fleet daily. This direct operational model link to the fleet was later exploited by LRAPP in both the model development and Fleet Operations support.

Since there was not enough computer capacity at FNWC, the system has to be operated in a multiuser environment. Operating in this way meant that when fleet support was not going on, the storage space could be automatically reprogrammed and the computer could be used by developers. This situation led to some very interesting discussions on priorities; nevertheless, the operation continued and progress was made.

In 1972, the Signal Processing Prediction System (SPPS) was operating. The SPPS was the precursor to the long-range system performance models. A contractor, Ocean Data Systems Incorporated (ODSI),[29] was developing perform-ance calculations to be done on the computer, and signal performance prediction was required. The environment was multidirectional, and the SPPS required careful investigation of the individual environmental elements.

The transmission loss model (see Chap. One) at that time was RP70, which was formerly the Wolff model in the 1969–1970 time frame. It took approximately 1 hour to do the calculations per radial. At a minimum, there were eight radials (45-degree sectors) for its long-distance calculations (1500 nautical miles). RP70 allowed range-dependent variation in both the sound speed profiles and the bathymetry, or water depth. RP70 was simply much too slow. Another model was required. This model was the Fast Acoustic Coherent Transmission (FACT), developed circa 1972. FACT was excellent for 200 miles, but could only use a single velocity profile and a flat bottom.[30] It was decided to run the RP70 overnight for many different cases, and then to relate the parameters that drove the RP70 to the continuous zone predictions they supplied to the VP community. The result was the Acoustic Sensor Range Prediction (ASRAP) Model.

Other transmission loss models began to be developed at FNWC.[31] At that time, the Fleet either made direct path detections or first convergence zones spaced about 30 miles from the source. (The exact distance depends primarily upon the characteristics of the sound speed profiles.) The fleet

---

[29] Directed by CAPT Wolff, who had retired from the U.S. Navy by then.

[30] There were no noise prediction models then. The first model developed was called the FACT Ambient Noise Model (FANM), and its development was a many-year effort. There were no signal processing, beam pattern, or shipping models. A shipping model, which is crucial as input to any low-frequency ambient noise model, was developed by Dr. Louis Solomon and Dr. Allen Barnes of Planning Systems Incorporated (1973), and was called the Historical Temporal Shipping Model (HITS). Mr. Ken Osborne developed the architecture for the components of the models. The components were built individually and in parallel with each other. Each component was considered to be part of the whole.

[31] Developers included Dr. Evelyn Ayres and CDR Tatro.

could detect nothing between convergence zones, so there were many relatively close regions where no submarines were detected.

The next major effort was the development of the ASW prediction areas. The specific ASW prediction area was that area where every-thing remained relatively constant. There were maps of the bathymetry, bottom loss, and major water mass boundaries. When these maps were overlaid upon one another, there were regions where everything remained relatively constant. Such a region became an ASW prediction area. It was the job of FNWC to perform all the acoustic performance predictions for each ASW prediction area. This information was calculated and delivered to the Fleet. Three frequencies were used for the calculations. On one day, one frequency was calculated for all the ASW prediction areas; on the next day, a second frequency; and on the next day, the third frequency. Then the process was repeated. The environment was sorted by latitude and longitude.

It was inevitable that Dr. Hersey and CAPT Wolff would collaborate, and CAPT Wolff would receive R&D money. Naval Weather Service funds were operational monies and could not be used for R&D efforts. Under various influences, CAPT Wolff became interested in underwater acoustics. Dr. Hersey, whose main interest was science, became involved in a real-time data network due to his relationship with CAPT Wolff. Further, Dr. Hersey and his subordinates also got substantial running time on the world's biggest and fastest computers. CDC was an ardent supporter because FNWC was the first to purchase their large computers and let them be tested by CDC personnel in FNWC spaces. The three all had much to gain, and the U.S. Navy profited.

SOSUS was now beginning to provide considerable information about transiting Soviet submarines. The Navy needed predictions about the capability for any SOSUS site for any beam to detect the Soviet submarines at different ranges. The calculation of transmission loss required the environment as input. That envi-ronmental data set was as difficult and time consuming to prepare as the model runs themselves.

After considerable argument and discussion about the needs of the users, it was decided to abandon detailed calculations. Approximation would be used whenever possible. This method of using approximations reduced the running time from 1 hour/radial to 1 minute/radial. The environmental data were put together from synoptic sound velocity profiles obtained from NAVOCEANO.

The cost to run the calculations for each SOSUS site was about $50,000, mostly due to the improved modeling techniques developed under

LRAPP sponsorship.[32] This amount was for only eight radials and for 1500 nautical miles from each site. Both the cost and the poor coverage were simply unacceptable. It was necessary to consider radials to be at 5° increments, giving 72 radials. Clearly, a different approach would have to be employed. Further, the need to switch to a climatological database that was in 5° sectors became apparent. The climatological database was then merged with the synoptic surface temperature. The result was an environmental data base that changed on a daily, even diurnal basis.

Due to the computing power at that time, it was considered that transmission loss could not be a ray trace. New models had to be developed. This meant that CAPT Wolff's model, RP70, and the basis for his beginning relationships with Dr. Hersey, had to be relegated to the museum. RP70 survived by making calculations for very short ranges. The results of PARKA, CHURCH ANCHOR, EASTLANT, and many other at-sea measurements gave substantial evidence that computer models could predict the behavior of ASW systems in general, and SOSUS bottom systems in particular. The models were used for new SOSUS site selections and new system designs and, in particular, they were applied to towed arrays.

LRAPP had a collection of people who worked on different elements of the computer programs. Mr. Charles (Chuck) Spofford invented the Antisubmarine Warfare Transmission Loss Model (ASTRAL). Dr. Ron Larsen of the Naval Ocean Systems Center (NOSC), San Diego, California, worked on the beamformer. The signal excess equation was calculated, and work was done on the elements of the equation; the signal excess calculation was known as the Automated Signal Excess Prediction System (ASEPS). Dr. Solomon and Dr. Barnes developed the HITS model. Mr. Spofford and others invented the Bottom Loss UpGrade (BLUG) model. Each person was an expert in the individual components. Mr. Osborne's group at ODSI did the integration.

Early LRAPP days saw constructive collaboration between the contractors and the Navy. The principal investigators were hand selected. This early collaboration did not last forever, and corporate competition eventually came into play. The Competition in Contracting Act made it impossible to just pick the best performer for a job without full and open competition. It is Mr. Osborne's opinion that such an effort could not be done today, given the current contracting climate. Although LRAPP had significant internal cooperation, there was some hostility between some organizations

---

[32] In the late 1970s the cost for calculations dropped from $50,000 to $1.50.

and some individuals. However, many people were devoted to retaining the professionalism of the program. Initially, there was a significant "not invented here" factor, but that ultimately was overcome.

Model Development

As FNWC grew and supported the fleet, which included providing the performance predictions for all the ASW systems throughout the world's oceans, a group devoted to the development of acoustic models directly focused on the needs of LRAPP evolved: the Acoustic Environmental Support Detachment (AESD), a detachment of ONR located at the Naval Research Laboratory (NRL). Its first officer-in-charge was Commander (CDR) Peter Tatro of ONR, who had originally worked for CAPT Wolff and with Dr. Aurand in the early days of LRAPP, and then founded and led AESD.

Computer models are the sand boxes and toys of scientists and engineers. Since the original and early work associated with a model requires very little money, almost everybody develops computer models for their own projects. This abundance of models leads to substantial difficulties between different models and modelers who have a tendency to compete for fame and fortune in the same area. The issue becomes serious when different models, focused on the same problem and driven by the same inputs, provide different predictions.

Thus, it became necessary to have model "bake-offs," which were sponsored by AESD. The issue was that models had to generate the same answers for the same inputs. These bake-offs were meetings where anybody could come and present their model results to the assembled community. The catch was that all the models presented were given the same set of input data, then the outputs of the different models were compared with each other, as well as with measurements. This process, while at times embarrassing for those models (and modelers) that made predictions that didn't compare well with data or with other models, was necessary. The process was done with reasonable diplomacy and tact, but regardless of the occasional embarrassments and obvious model failures, proceeded. The only thing that really counted was the ability of the computer modes to have independent agreement with the measurements.

There were other issues, too. Speed and model size, were two measures. Since the ultimate user of the models and their predictions provided by FNWC was the operating fleet, the correct predictions were very important. But, the computer models had to run sufficiently fast, and also had to be capable of being run on existing computers of that time.

The Navy uses many systems. Their optimum performance depends upon consistent and correct predictions from Navy Standard models. Dr. Hersey was the modeling coordinator of the U.S. Navy for oceanographic models. He was responsible for development and certification of Navy Oceanographic Standard models. Existing systems used the computer models developed by the Navy laboratories. Unfortunately systems using different but ostensibly similar computer models frequently were provided different predictions for the same problem. For example, there were two high-speed, short-range transmission loss models that used a single velocity profile and a flat bottom: FACT and RAYMODE. While both computer models were based upon the same physical assumptions, and used the same forms of available input data, they provided, sometimes, different, incompatible predictions. Which model was correct? How would the operators at sea know which answer to choose if they had two different answers? The fleet, who needed to use the model predictions, would simply threw up their hands in disgust. It was absolutely imperative that computer models used by the fleet provided the same answers for the same input data. Eventually, this led to the concept of competition between two or more similar models. This led to the selection of one and one model, only, which became certified as the Navy Standard Model to be used for a specific set of circumstances. This model provided only one answer, and that was the answer used by the fleet.

What is a Navy Standard Model?

A Navy Standard model is a model that has been certified by the U.S. Navy as being "the" model that is to be used by the Navy and its laboratories. No model is declared to be the standard until after it has undergone rigorous review and approval.

FACT eventually became the interim Navy Standard model for transmission loss. Later, the Parabolic Equation Model (PE), developed by Dr. Fred Tappert, became the Navy Standard model for transmission loss. PE evolved from its original form where it was developed and used as a model to predict radar performance. Initially, PE couldn't handle upslope calculations, but as it was further developed, that problem was solved. PE used a triangular scheme developed for interpolating velocity profiles by Dr. John Hanna that worked very well.

An interesting issue grew within the surveillance community. Substantial conflicts arose between the Navy laboratories and Bell Laboratories. Within the Navy, each laboratory was funded by the parent Systems Command (SYSCOM). At that time, Naval Air Development Center

(NADC) was funded by NAVAIR; Naval Undersea Systems Center (NUSC) was funded by NAVSEA; and Naval Operations System Center (NOSC) was funded by NAVELEX. These laboratories received 6.2 money. AT&T worked on surveillance, but there was no SYSCOM for surveillance.

With the continued success of SOSUS, the Navy laboratories began to be interested in supporting SOSUS, which really had been developed by AT&T. As a part of OPNAV, OP-095 was worried because there was no real technical organization that could or would challenge AT&T. However, LRAPP provided that opportunity for real scientific competition. While competition between the working scientists and engineers was relatively pleasant and cordial, there was real tension between the leaders of these organizations, and the primary reason was the financial support provided by the SOSUS system, OP-095, and PME-124.

All the data were received from Western Electric. Both Western Electric and Bell Laboratories were wholly owned subsidiaries of AT&T. It was decided that future SOSUS system detection arrays would be placed in the sound channel, as well as on the bottom. That meant that it was necessary to find seamounts (underwater mountains) that rose from the sea floor into the sound channel. Finding such places involved reviewing the bathymetry to find seamounts that intruded into the sound channel. Just before the sound arrived at the hydrophone array, it interacted with the bottom, or in this case, the side of the seamount. This effect caused a major effort to focus on sound interaction with the bottom. The data used for development of bottom interaction models came from the old SQS-26 worldwide survey. That data, painfully and expensively gathered in support of the SQS-26 system by Alpine Geophysical over many years for NAVOCEANO, still existed. Mr. Spofford and others used those data to develop the Bottom Loss Upgrade (BLUG) model that replaced the ancient curves (called the Bassett-Wolff bottom loss curves) developed by Mr. Charles Bassett and CAPT Paul Wolff.

AESD supported the SYSCOMs and delivered results of transmission loss and ambient noise calculations. One other major effort of AESD, in addition to supporting the development of different types of models, was the preparation of common data sets necessary for use in Requests for Proposal (RFP). Big companies that design and manufacture hardware systems for use by the military always attempt to demonstrate their systems in the best possible light. However, it is the responsibility of the government and its representatives to make sure that these expensive systems will operate as advertised in a variety of different environmental and operational scenarios.

For that reason, when RFPs were issued by the government and sent to the industrial community, data sets were prepared and provided to each interested contractor. They were required to show how their system would perform under each of the environmental and operational scenarios. This information would allow comparisons between the capabilities of the competing systems. This approach, so standard today, was met with considerable reluctance by the industrial community. They were accustomed to present their systems in a favorable light and leave it to the Navy to decide how the systems would work in not-so-favorable conditions. When this procedure was implemented the first time, in support of the Moored Surveillance System (MSS), at least one contractor objected strenuously, raising the objection to the Secretary of the Navy. After listening to the arguments in an emotionally charged meeting where AESD was represented by CDR Tatro, the Secretary wisely decided to support the process. It was a precursor for the procedure that is followed to this day.

Most people focus on the at-sea experiments, which were the most expensive part of the LRAPP effort. These experiments get the most publicity and can be extremely exciting, as well as interesting. But they take data in one general area, at one time. These data cannot be changed to be applicable for another time and place; however, those data can be used to validate and verify the predictive capability of the computer model.

Computer models have the power to live past the experiment. A system can also be built using only models. Not only do the models subject the system to substantial tests in different areas under different environmental conditions, they also are critical in ascertaining how the actual physical systems should be deployed. Finally, models allow the results of either experiments or operations to be interpreted and explained. Thus, in the final analysis, it was the suite of computer models that was the ultimate product of LRAPP.

CHAPTER Ten
Exercises

General Comments

The most obvious activity of LRAPP was the unceasing planning, implementation, analysis, and reporting of field exercises. These exercises were not unique in the annals of the U.S. Navy, but the way they were conducted was unusual. Historically, most naval field exercises were run by universities or navy laboratories and were generally supported by basic research (6.1) or exploratory development (6.2) funds. As such, they had little interaction with fleet units. Further, they were generally conducted without a sense of urgency. Experiments were planned, measurements made, and results eventually analyzed and finally reported in erudite professional journals. The results were not readily available to the senior government and naval officials, nor were their implications easily understood.

The LRAPP field exercises were radically different. They were supported by advanced development (6.3) funds and had close interactions with fleet units. They were designed to produce results that had possible national repercussions, and their visibility from senior naval and government officials was high. In the course of the LRAPP, approximately 96 exercises[33] took place over the life of the program. This number of exercises has no precedent in naval/scientific investigations.

Each exercise took approximately 18 months from conception to conclusion of the actual event. To some extent, this period of time was due to the direct result of Dr. Gaul's leadership. He insisted that the exercise be planned in meticulous detail. Not only was this important, it was also mandatory. Dr. Gaul had an unusual procedure that was generally not

---

[33] The number is approximate because some exercises had different phases and were sometimes counted as separate exercises, and sometimes as part of the same exercise. Further, at least one exercise, BEARING STAKE, was not paid for with LRAPP funds, but it was directed by Dr. Robert Gardner using all LRAPP personnel and support contractors. BEARING STAKE belongs with the LRAPP list of exercises for all intents and purposes.

followed by any other manager associated with at-sea experiments: he divided the funds into two separate parts, roughly equal in size. One part was for the actual expenses associated with the exercise. The other part was for the data analysis and reporting.

As a rule, other managers did not use this demanding procedure. The result frequently was that most if not all of the funds that their programs received for ocean exercises were spent on the exercise itself. Regardless of the quality of data, there frequently were no funds left to perform the analysis. Dr. Gaul felt, and was strongly supported by many members of the scientific community, that unanalyzed data and data unreported in a timely fashion was equivalent to having no data.

The process of running field experiments first required specification of the objectives of the experiment. What needed to be measured? What equipment would be used? Where would the exercise take place? What platforms (ships and airplanes) would be required for support? Once these questions were answered and agreed to by the senior naval managers, the detailed planning would begin.

It is useful, as well as necessary, to comment on the scale of these experiments. As an example, consider the CHURCH ANCHOR exercise. It was run in 1973 in the Northeast Pacific. It was roughly bounded by the west coast of the United States, the International Date Line, Hawaii, and Alaska. This area comprises approximately one-fourth of the Pacific Ocean. Most of the measurements were made along the experiment baseline, which was 143° 30' W longitude for a distance of approximately 2000 nautical miles. The implication of this geography was that many ships that took part in the experiment never saw one another. They left for participation in the exercise from different ports at different times, did their assigned tasks at sea at different positions according to a rigid time schedule, and left the exercise for their individual destinations once their jobs were complete. The airplanes used in the exercise took off from military air bases throughout the North Pacific region at different times, flew different tracks, and returned to their designated bases at the end of their missions.

The complexity of this effort was directed by the chief scientist or the technical director, and was coordinated by him (or her) from a convenient place. There was communication among all of the exercise elements using radio and military messages.

Be Prepared!

Mechanical and electrical systems fail, so during at-sea exercises, every platform had to have the capability for in situ repairs.

During the CHURCH ANCHOR exercise, I was the Chief Scientist on the source ship, the M/V *Mediterranean Seal*, a 155-foot LOA supply ship. The source was a derivative of a land-based seismic source, Vibroseis. Two elliptical heads would be hydraulically driven to generate a controllable sound source. Operating parameters were impressive: operating depth of 500 feet, frequency range of 10 Hz and higher, and source levels to 209 dB. The energy was supplied with large diesels, Sundstrand pumps, complicated valving, and multiple parallel filter banks to the 3-micron level. As soon as the system was made operational, it was on its way downhill. The question was, when would it fail?

We had spared everything that we considered critical. Halfway through the exercise, we began to run low of Sundstrand parts. I advised Exercise Control in Honolulu of the problem. Shortly afterward, we were notified of an airdrop from a P-3 aircraft. The day of the drop, the P-3 leveled off at 150 feet of altitude and 200 knots, and dropped a 30-gallon metal container from the plane. We retrieved the package, but nothing was usable. I notified Honolulu that the resupply effort was a total failure. Honolulu said they were surprised: they conducted a test with the package by dropping it from a three-story building! I suggested that next time, they try a fifteen-story building traveling at 200 knots, or maybe a parachute.

Someone at Exercise Command Center was thinking. A second set of necessary spares was given to the M/S *Desteiguer* when she left San Francisco. She was one of the ships in the CHURCH ANCHOR Exercise. We rendezvoused with her five days later and obtained the necessary spares. Except for exceptionally bad weather and Force 10 seas, the Vibroseis was operational during CHURCH ANCHOR.

Claude P. Brancart

Time was the only common element that allowed the integration of all the data after the exercise was complete. All activities were governed by Greenwich Mean Time (GMT) or using the military designation, Zulu Time. Accurate time coordination was mandatory. Ships in one part of the Pacific did their assigned tasks at the same time other ships did theirs. The individual platforms had a senior scientist on board who regularly informed the technical director of his or her activities, difficulties, etc. If there was a problem, then it was up to the technical director to make changes in the exercise schedule. Such changes were initiated only under the most compelling conditions. There were several ships, many aircraft, and dozens of people whose sole point of contact was the technical director and his support staff at the home station.

At the end of the experiment a meeting was held by most if not all of the participants at a convenient place (usually an international airport close at hand). This meeting, called the Hot Wash-Up, allowed all the data to be transferred to the control of the technical director and his analysis support staff. All pertinent information was discussed, and the general and specific details of the individual platforms were told to all the people attending. These meetings, while informal and usually punctuated by "sea stories" of the happenings during the exercise, were crucial. Each senior scientist gave a preliminary report on the quality of measured data, and it was immediately clear to all whether the experiment was a success or failure. *In the 96experiments run by LRAPP over 25 years, there was never a major failure.*[34] The participants were forced by Dr. Gaul to have plans, contingency plans, and even further contingency plans to make sure that data would be obtained and that the objectives of the experiment could be met. The cost of such experiments, depending upon the size of the exercise, was generally measured in millions of dollars. A failure of a critical element, measurement device, or platform would mean that the entire experiment produced no data. The money would have been spent, and there would be no results. The responsibility for assuring success in obtaining the data was felt by the entire scientific community. It was frequently stated that they were working for the United States taxpayers and success in obtaining data was a personal responsibility.

Dr. Gaul also insisted that the modeling community step up and be counted. While the glamour, whatever there is, generally resides with the people who go to sea, take the measurements, and have wonderful stories to tell after they return, the real fundamental results that LRAPP delivered

---

[34] Some people consider SEA SPIDER and TEST BED to be major failures.

to the operating fleet were predictive models. At any given moment, the fleet has only certain information available to it. This information translates into knowledge of the environment (summer or winter, which ocean, stormy or clear, etc.) and some intelligence information (e.g., a submarine left Soviet home waters and historically followed some general path to its operating area).

In addition, only certain types of equipment were part of the fleet's operating inventory. They needed computer models that would use all the existing data (and no more), and would provide the fleet with predictions of what would be their best tactics, given the constraints of equipment, geography, and anything else, to provide the highest probability of detecting and then tracking the Soviet submarines. Therefore, Dr. Gaul insisted on the "Mason Jar" approach. When an exercise was finally planned, i.e., it was decided what was going to be measured, at which fre-quencies, at which depths, etc., etc., etc., the modeling community had to generate predictions for each and every measurement. These predictions would be made before the experiment was begun; the results were placed in a sealed Mason jar and then compared with the actual measurements after the exercise was over, the data were analyzed, and the data put in a suitable format that would allow comparison to predictions.

The modeling community was reluctant to do this, but Dr. Gaul prevailed. The knowledge that comparisons would be made between actual measurements and model predictions produced several interesting results. First, the measurement people

> One of the destroyer escorts had the misfortune to have all its electronic navigation systems break down a day and a half before setting sail for a coordinated multiship task. It soon transpired that none of the crew members knew, or perhaps remembered, how to navigate with a sextant.
>
> The civilian observer aboard, Don Sullivan, was, however, an avid amateur yachtsman and had taught the Coast Guard's "Power Squadron" course on celestial navigation. He verified that the ship actually carried a sextant, so he took a few officers to a spot on the deck out of sight from the passers-by on the pier, and taught them how to navigate without electronic signals.
>
> The issue became moot; although the ship left port without any of its electronic navigation working, the equipment was up and running before they reached their first station.

knew that their results were going to be considered as fundamental truth. This knowledge increased their concern with accuracy, multiple measurements, contingencies, and many other techniques that would provide a database that was as precise as possible.

Second, the modelers really focused on the results of their models. When the predictions of measured re-sults were presented in a semipublic forum (highly classified, but all the mem-bers of LRAPP were there) and found to be substantially different from the meas-urements, it became neces-sary for the modelers to explain why there were differences. It was a most unpleasant time, and modelers simply didn't like it. Therefore, considerable care and effort was used to increase the capabilities of the models.

Third, the modelers came to realize that not only was predictive accuracy required, but also the calculations must be obtained in very short periods of time. The fleet has hundreds of operating units, and they operate world wide for 24 hours a day, 7 days a week. At this time of computer development, all the calculations were to be performed at FNWC and the results transmitted to the operating fleet on a daily, and frequently hourly, schedule. Thus, the modelers need to develop fast and accurate models.

Due to the small delay between the end of the experiment and the Hot Wash-Up, there was some time for preliminary major results to be considered, analyzed, and discussed. These results, preliminary and incomplete as they were, served as the basis for the presentations to the senior naval and civilian command authorities. This process of making these presentations was generally referred to as the Road Show, and Dr. Gaul and a host of people who had actually done the work (generally the technical director, the chief scientist, and the appropriate senior scientists) presented the results at several dozen (!) meetings.

Of particular importance were the presentations to the Fleet. In many cases, LRAPP was heavily involved with the fleet during the course of the experiment and worked on problems of direct and immediate concern to them. The scientific results of the experiments were placed in the context of operations. LRAPP may have been the only program that regularly worked with the fleet and made their results known to them in the period immediately after the conclusion of the experiment. This created a very strong bond between the Fleet and LRAPP.

The work relationship between LRAPP and the operating forces in Third Fleet (E. Pacific), Seventh Fleet (W. Pacific), Sixth Fleet (Mediterranean) and Second Fleet (Atlantic) was carefully nurtured and was quite strong at all levels. From the earliest days of LRAPP, there was always a naval officer, and frequently two, who were part of the LRAPP staff. The officers served as the liaison between LRAPP and the U.S. Navy.

The Navy consists of officers and enlisted personnel who consider themselves professionals, and they are. They have their own language, culture, history, and procedures for operating. They have procedures for everything and, as a general rule, have great reluctance to do things that are not part of their training and practice. They regard civilians (anyone who is not military, and particularly not Navy) as people to be tolerated, and perhaps to provide amusement from time to time, but certainly not capable of helping the operational Navy solve its day-to-day operational problems. They are not entirely wrong in this view, since civilians generally don't know or understand the Navy culture and way of doing things, and as such are frequently misunderstood.

LRAPP solved this problem in several ways. The first was to have professional naval officers as part of their staff. It was quite instructive to see commanders and lieutenant commanders accompany Dr. Gaul and others when they went for meetings with the Commander, Sixth Fleet (generally a vice admiral). Hierarchies exist in all organizations, and perhaps a little more in the military than in nonmilitary organizations. Within the Navy culture, lieutenant commanders don't usually talk directly to vice admirals, and usually don't attend working meetings with them. Yet LRAPP marched in and part of its staff was a lieutenant commander.

In fact, the relationship between LRAPP and pertinent naval commands prospered because of this. Dr. Gaul and the senior civilians would discuss the details of the experiment, and a general agreement would be reached between the scientific and naval leaders. Dr. Gaul would then indicate that the lieutenant commander would be his point of contact for all matters relating to interactions with the Navy. The admiral would then feel at ease because he knew that a professional naval officer would interact with all the elements of his command, observe the proper protocols, send the proper messages, understand the requirements of Navy operations, etc. And it worked. All the naval officers associated with LRAPP recognized their highly unusual positions and considered it to be an unusual trust. Each of them earned the respect of their naval superiors, and each was promoted regularly. Further, the operating units of the fleet, both at sea and ashore were extremely responsive to requests from the LRAPP naval officer for two reasons:

- They were naval officers and clearly knew what they were doing and knew how to ask for things in the navy way.

- It was clear the this particular naval officer, while only a lieutenant commander, had direct access to the fleet commander. This access was highly unusual, but it was understood by one and all that this was the only way to assure the highest probability of mission success.

LRAPP assured its success with the Navy in one other way. The LRAPP community in general recognized that they would be, albeit briefly, on intimate terms with the operating fleet: LRAPP would use naval equipment as well as their own, and they also would use the Navy communications system. LRAPP would become part of the Navy in an odd way. Everybody who was a member of LRAPP came to the conclusion that they would make themselves acceptable to the Navy at all levels. So, everyone did things "the Navy Way." In addition to doing things the Navy way, LRAPP also showed the naval personnel what they were doing and how their results were specifically focused on the problem of finding Soviet submarines. After initial skepticism (in each and every exercise), LRAPP personnel found that the naval personnel respected them and treated them as colleagues with special capabilities that could help them in their day-to-day job: defeating the Soviet submarine force.

Field Exercises

The 96 field exercises were arranged roughly into two separate groups: scientific and operational. The scientific exercises were predominant by far.

LRAPP went to all the oceans in the Northern Hemisphere[35] and the Indian Ocean (BEARING STAKE). Each exercise had specific and detailed objectives. As a rule, these objectives involved the measurement of certain ocean characteristics: temperature, salinity, and speed of sound at various points throughout the area as a function of depth from the sea surface to the bottom, and bottom measurements). It also entailed measurement of transmission loss and ambient noise, as well as the components of ambient noise (merchant ship locations). The

---

[35] The location of all the LRAPP exercises are on a chart of the world, with a detailed chart for the Mediterranean in Appendix A. The numbers on the charts correspond to the location of the exercises, which are also listed by number.

transmission loss measurements were made by dropping SUS[36] charges. These charges were thrown overboard from ships or dropped from aircraft. They exploded at preset depths (shallow at 60 feet and deep at 300 feet). These depths were selected because the target (submarines) operates relatively near the ocean surface.

A Case of Beer

Roy Gaul always had a penchant for arriving late to an exercise planning meeting, rearranging the seating patterns, and then doing most of the talking. At one meeting, Bud Williams, who was responsible for integrating the exercise plan, got me to the side and said, "If you can keep Roy from talking at the next meeting, I'll buy you a case of Coors!"

I was horrified, of course, at Bud's suggestion. Nevertheless, when Roy and I got back to ONR, I told him of the bet and said I would split the case of Coors with him if he could maintain his silence at the next meeting. He, being the sport he was, readily agreed.

At the next meeting, in a room full of people, Roy kept his silence while Bud and I conducted the meeting. This went on over the course of two hours plus and then I adjourned the formal meeting. Bud, at the other end of the table, reacted by letting his jaw drop and exclaiming, "Glynn, you s.o.b., you set me up!" Next time out, both Roy and I had a case of Coors awaiting us in our hotel rooms.

There are many, many more stories about the colorful characters and events with the LRAPP entourage, but I will save those for my "book"! I will end this by saying it was a fast, sometimes grueling, but always fascinating, two years on CHURCH STROKE and I came out of it, if nothing else, well traveled!

---

[36] SUS is the acronym for source underwater signal—small explosive charges set to explode at selected depths.

Although submarines can operate shallow, deep or very deep (close to 1000 feet), their position can best be characterized as either above the frequently existing sharp break in the sound speed profile near the ocean surface (shallow) or below it (deep).

The variation in the transmission loss as a function of the source depth (the submarine or the SUS charge) is really quite small, except for the difference between being above the thermocline break or below it. The speed of sound above the thermocline break and below the surface is strongly affected by surface conditions (i.e., winter or summer, stormy or not, windy or calm). Additionally, the shallow sound speed profile is different for each ocean. Below the thermocline break the sound speed profile is much more dependent upon the local salinity (almost invariant), the temperature, and the pressure (strictly a function of the depth below the surface).

The receiving system was located at different depths, depending upon the ocean characteristics (depth of water, sound speed structure). The receivers were of different types. In the early days, the receivers were omnidirectional and were arranged at preset depths from near the ocean surface to almost on the bottom. These devices measured both ambient noise and the SUS signal arrivals. This measurement allowed the calculation of transmission loss. Another type of receiver was a towed array, which was a directional receiver. It was more sensitive looking in one direction than another.

While the basic focus of the scientific exercises was to gather scientific data to support model validation, characterization of the environment, and in general get to know more about the ocean, once in a while Soviet submarines wandered into the area and became unwitting participants in the exercise. The fact that they were detected in real time (generally with towed arrays since the measured data was seen immediately on board the towing ship) was relayed to appropriate naval operational centers using proper Navy communication procedures. The Navy invariably became excited about this development, and sometimes wanted us to abandon the experiments and follow the submarines. This departure from the scientific experiment was never made, since other exercises were specifically designed to follow the Soviets. Initially, this caused some irritation, but the Navy understood that we were playing for the long term and grudgingly admitted that we were right not to be responsive to operational circumstances.

The operational exercises had a different focus. We went to sea with our receiving equipment, usually towed arrays, specifically to find Soviet submarines. And we were almost always successful. The fleet was grateful and sometimes extremely surprised when we told them about a Soviet submarine of which they had no knowledge. Such a discovery resulted in the intelligence community telling us that couldn't be; confirmation was required. They were right, of course, in expressing skepticism. However, when we could direct a P3-Orion aircraft to a site where we contended the Soviet submarine was operating, and pictures were taken when it was on the surface, even the most vocal and skeptical of the critics were generally silenced.

These fleet support exercises, where, in effect, we became an adjunct to the operating fleet, were initially looked upon with disbelief by the operators themselves, but when we were successful and shared our results and methods with them in the appropriate Navy manner, they first became grudging supporters, then strong advocates of LRAPP. These exercises were conducted throughout the world, and frequently we encountered the same naval personnel over and over again. They went out of their way to welcome us as fellow brothers-in-arms, and warm and close relationships were maintained without any rancor or strain between LRAPP and professional Navy.

Of the many exercises that were run by LRAPP, several were perhaps a little more important than others. The reasons for this importance are highlighted here. These exercises are PARKA.[37] EASTLANT, CHURCH ANCHOR, SQUARE DEAL, TASSRAP, MASWAP (Mediterranean ASW Augmentation Program), BIG DIPPER, CHURCH STROKE, and BEARING STAKE.

PARKA

PARKA was located in the Northeast Pacific and is detailed in Chapter Seven. The exercise was enormous and was the first exercise that was part of the LRAPP program. *PARKA was probably the best experiment ever run in terms of the amount of data actually taken, the care and precision with which the data were taken, and in setting the standards for all future at-sea exercises.* Transmission loss was measured, but strangely ambient noise was not. All the techniques and platforms were tried during this test except sighting (rigging) of surface ships for ambient noise prediction purposes.

---

[37] Some of these exercises had different phases or components. When the name of the exercise is used here, it refers to all phases of the exercise.

How I Learned Who Was to be the Chief Scientist

One fine day I set off in a plane for Washington in response to a summons from Roy. At midmorning I found myself at ONR in a room by myself, waiting. Roy soon dashed in, put his hand on my arm, and led me rapidly to a nearby small auditorium, saying "Come on, there are some people here I want you to meet."

Among the dozen or two people in the room were a few whom I recognized and many more who were strangers. I prepared myself to shake hands and smile politely with those Roy would introduce me to. Instead, he announced in a voice loud enough to call the whole room to attention, "I want all of you to meet the chief scientist for this exercise, Tobey Raisbeck." In a few sentences he outlined why it was important for me to confer both one at a time and en masse with the people in the room, who turned out to be the leaders of various groups of people who would be collaborating with ONR by carrying out experiments and field exercises. He closed his introduction saying to me, "Why don't you tell us all what you plan to do and how you propose to go about it."

By the time I recovered enough composure to say anything at all, it was too late to do anything but move forward. And so, even though I did not know who most of these people were or what they expected to be doing in the exercise, I presented a plan.

Over the next year and a half, a number of important measurements were carried out. One of these measurements firmly demonstrated the important fact that low-frequency ambient noise in many places is far from uniformly omnidirectional. In spite of its chaotic beginning, the exercise paid off.

Tobey Raisbeck

EASTLANT

From some perspectives, EASTLANT might be considered to be the most important exercise that LRAPP conducted. The situation of the U.S. Navy and AT&T disagreeing about the implantation of SOSUS arrays in the eastern Atlantic has been discussed, but certain facts bear repeating here.

The Navy was frantic, since there were no indications and warnings (I&W) of Soviet submarine activity in the open ocean. The ASW problem, always difficult and thought perhaps impossible to solve, was even worse without I&W. It was the first element in the prosecution of an enemy submarine.

AT&T, or at least one part of AT&T, insisted that the ambient noise levels were so high that the S/N would be so low that the range for possible detection of submarines would be too small to provide any intelligence of value. Dr. Hersey and Dr. Gaul, supported by members of the LRAPP community, insisted that the ambient noise in the part of the frequency spectrum of interest (lower frequencies about 50 Hz) was directional. *That one piece of information, if true, would radically affect the methods which would be used to search for submarines.*

VADM Shear had already delayed the implantation of the SOSUS arrays for one year. He couldn't wait any longer. Either he would cancel the implantation and have no solution, or, he would let Dr. Gaul and his scientists see if a possible solution could be found. It was the good fortune for the nation, ASW, and LRAPP that VADM Shear decided to let Dr. Gaul try to prove his theories.

The results clearly demonstrated, even with the relatively crude equipment of the moment (ITASS arrays towed by destroyer escorts), that the ambient noise was directional, showing 10–15 dB variation with azimuth. This result proved to OPNAV and to the ASW military and civilian leadership that civilian scientists focused on ASW and working hand-in-glove with the fleet might be able to provide the scientific tools that the operational Navy was lacking. In essence, not only did EASTLANT provide an answer that saved the fundamental basis for the SOSUS system, it also established LRAPP's credibility, which was never lost for the rest of the life of LRAPP.

What is a Thermocline?

The *thermocline* is the characterization how the water temperature varies with depth.

CHURCH ANCHOR AND SQUARE DEAL

Both were gigantic exercises run almost simultaneously. CHURCH ANCHOR was run in the Northeast Pacific, and SQUARE DEAL was run in the eastern Atlantic. Measurements were made of transmission loss and ambient noise. In addition, ships were counted and their spatial distributions were mapped. *This process became a standard part of all future exercises,* since it was the distribution of these surface noise sources that contributed to the directionality and levels of the ambient noise in the frequency band of interest.

It was during these exercises that the final methodology of defining the experiment, working out the details on a master timeline,[38] and involving the development of a data analysis plan before the exercise ever went to sea were accomplished. All future exercises followed the same proven successful procedures.

TASSRAP

The TASSRAP exercise used towed arrays in the Mediterranean for the first time in a manner dictated by the environment. Close relationships were established with the Sixth Fleet, as well as a tenuous working relationship between the scientific community and the operating fleet, with cautious respect on both sides.

MASWAP

MASWAP (Mediterranean ASW Augmentation Program) continued the relationships begun in TASSRAP, and resulting in close working relationships between the two communities.

BIG DIPPER

BIG DIPPER was another Mediterranean exercise. It used the mid-frequency array (MFA). Up until that time, the Navy and the scientific community had focused on arrays that were sized for 50 Hz. Based upon

---

[38] Timelines for the exercises are given in Appendix B; exercise photographs are located in Appendix C. Appendix D is a glossary. Appendix E is the members of the LRAPP community, and Appendix F is a technical discussion of the foundations of Underwater Acoustics.

scientific measurements made by LRAPP, it was decided to build an array that was based upon the sixth harmonic of the fundamental power frequency of 50 Hz, which was 300 Hz. Although in principal the transmission loss was lower at 50 Hertz than for 300 Hz, in reality—in the Mediterranean—it didn't matter because it is a small sea, certainly when compared to the Atlantic and the Pacific. Further, the noise would be much lower at 300 Hz than 50 Hz. The ambient noise at 50 Hz is higher in the Mediterranean than any other ocean area on earth.

The MFA was first tried during BIG DIPPER. The Sixth Fleet gave an award for the best performance by a military unit working on ASW. This award was highly regarded by the operating units. The award, known as the Hook 'Em Award, which was in the form of a submarine being hooked by a giant fishhook, was classified at the SECRET level. During BIG DIPPER, the LRAPP crew, using the MFA, located several Soviet submarines whose presence was not known either by intelligence or by any military ASW units. When LRAPP reported that a Soviet submarine was in the Mediterranean in a place where no one else had made any detections or even suspected the presence of a Soviet submarine operating in the Mediterranean, this signal was met with disbelief. However, after some discussion of the data,

CHURCH STROKE Denizens

I was newly arrived to LRAPP staff when soon thereafter the planning began for the CHURCH STROKE series of exercises. I volunteered to Dr. Gaul to be LRAPP's Pro-gram Manager, given that there seemed to be a vacancy at the time (my first mistake!).

Now, I had been around some of the LRAPP meetings in the past and knew Roy always did things in a big way accompanied by large crowds of people—many of them whom I knew at the time. However, I was destined to meet an even larger number, with diverse temperaments, and of wide geographical distribution.

My first experiences of exercise planning, Roy-Gaul-style, took place in the Xonics facilities in Los Angeles with people flying in from all directions: formal sessions, rump sessions, breakfast-lunch-dinner-and-way-into-the-night sessions.

Glynn Lewis

a decision was made to bring additional military forces into play, and the LRAPP detection was confirmed with a photograph of the Soviet submarine caught on the surface. For detecting this unknown, unsuspected Soviet submarine, the Hook 'Em Award was presented to LRAPP. As far as is known, this is the only time that the award was presented to a civilian crew using nonmilitary equipment.

CHURCH STROKE

CHURCH STROKE was a major exercise consisting of several different phases. It lasted for over one year. It was in the Pacific. During CHURCH STROKE the Large Aperture Marine Basic Data Array (LAMBDA) was used. *This exercise was the first time that LRAPP used a research array, LAMBDA, to augment its data collection.* In addition, the close relationship between the LRAPP and the operating fleet (in this case it was the Third Fleet) continued, and they began to depend upon each other. LRAPP brought its scientific and research tools to bear upon military operational problems. In effect, this exercise began the strong coordinated ASW operations that began to be used throughout the world by the U.S. Navy.

BEARING STAKE

> No database on the environmental and acoustic charac-teristics of the Indian Ocean is even remotely comparable to the BEARING STAKE data.

BEARING STAKE was not a LRAPP exercise in that no LRAPP funds were involved. The exercise was sponsored and paid for by PME-124, who was commanded by then-CAPT Dempster Jackson.[39] The office of PME-124 was responsible for managing the SOSUS system. The exercise was held in the Indian Ocean during the International Geophysical Year. It was on a scale similar to CHURCH ANCHOR. The technical director was Dr. Robert Gardner, and due to his experience as part of the LRAPP team, he ran the

---

[39] He was later promoted to Rear Admiral and is now deceased.

exercise using all the members of the LRAPP community[40] and used the techniques that were employed so successfully as part of the LRAPP efforts.

Mediterranean Operations

Mediterranean Operations refers to a series of combined operational fleet efforts that seemed similar to a LRAPP exercise, except that no scientific measurements were made. This exercise was strictly a fleet operation that focused on finding and prosecuting Soviet submarines. However, just as EASTLANT demonstrated the utility of the scientific community in support of the ASW community and established their credibility, this particular short series of events demonstrated to the Fleet that application in an operational scenario of what LRAPP had developed—models and tactics for fleet use based upon model predictions and understanding of the effects of the environment on submarine detections—led to success.

[40] Appendix E is a listing of members of the LRAPP community.

CHAPTER Eleven
Arctic Operations

General Comments[41]

By the early 1980s, Soviet strategic submarine operations began to change from forward deployments off the coasts of the United States and in the Caribbean, to patrols that did not leave the waters near their submarine bases on the Kola Peninsula and the Seas of Okhotsk and Japan. The United States responded to these developments by promulgating a plan called the Maritime Strategy. This plan called for an immediate, large-scale naval response to Soviet aggression in Europe that would take the war to "forward areas," the euphemism for waters near the Soviet Union where their warships, including submarines, operate.

Since Soviet strategic submarines then operated in "bastions," the Maritime Strategy required that the United States be able to "penetrate the bastions" by conducting a strategic ASW campaign in the bastion areas. This strategy established the requirement for conducting large-scale ASW in polar conditions, including near the edge of the polar ice pack, the so-called marginal ice zone, and under the ice in the deep water of the Arctic Basin.

The areas near and under the ice are environments different from those where submarine and antisubmarine operations were usually conducted in the early 1980s. Most notable is the constant presence of floating sea ice, often completely covering the ocean surface for many square miles. The presence of sea ice complicates ASW in many ways, especially by denying free movement to surface ships and by denying access to the water to most air-deployed sensors and weapons. The ice cover thus provides added protection to the submarine from ASW forces.

---

[41] I am indebted to Nancy Bedford of ARL:UT for this material.

Program Objectives

In response to the requirement for an essentially new warfighting capability, the CNO's Arctic Initiative was promulgated as formal policy when Navy Instruction OPNAVINST S3470.6 was issued in March 1983. The elements of the CNO Arctic Initiative are listed.

- torpedo improvement
- surveillance
- sonobuoy
- 688 submarine class backfit
- under-ice piloting sonar
- submarine sonar
- submarine communications
- acoustic countermeasure evaluations
- offensive mine performance
- NAASW (Non Acoustic ASW)
- structural stability
- search and rescue
- Arctic Submarine Laboratory support
- environmental support

The last of the 14 priorities, environmental support, was seen as an umbrella task that applied to each of the other 13 issues.

The Oceanographer of the Navy, then OP 952, responded to the CNO's initiative with a new Arctic Initiative in the Program Objectives Memorandum (POM) 85 cycle. This program set out to establish the environmental performance baseline and to improve performance for torpedo, SSN, surveillance, and air ASW capabilities in the Arctic. As part of this effort the AEAS[42] Program developed and issued a specific plan of action entitled "AEAS Arctic Environmental Program, POM-85 Initiative, Plan of Action and Milestones." The primary goal of the program was to provide sufficient

---

[42] The new name of LRAPP. AEAS stands for ASW Environmental Acoustic Support

understanding and realistic environmental databases to allow the selection of logical alternatives in sensor design, signal processing, and system configuration for Arctic operations. To reach this goal, measurements, modeling, and analysis objectives were established.

Measurement objectives included the design of and participation in field exercises to collect data needed to fill data voids. These voids included environmental and acoustic data at almost all locations in the Arctic. Acoustic data was needed to address the effects of the Arctic environment on ASW systems performance, including seasonal ambient noise levels and directionality, propagation loss, signal arrival structure, and signal coherence. A large part of the measurements program included the design and acquisition of instrumentation to collect both the environmental and the acoustic data.

The modeling objective was to provide an ice-capable modeling system with acoustic propagation models and Arctic environmental databases. Accurate acoustic modeling was needed for interpreting measurements products and for predicting performance of current and future ASW systems.

The analysis objective was to provide ASW guidance and site/system assessment products to the surveillance, submarine, and air ASW communities, as well as to the R&D community. The approach was to obtain these products by combining the systems and threat characteristics with the results of the transmission loss and ambient noise measurements where available, using modeled propagation and databases to address locations, frequencies, and depths for which measurements were not available.

Arctic Knowledge Prior to the AEAS Arctic Program

Small springtime ice camps had been set up in the Arctic for many years prior to the AEAS Arctic Program, and sonobuoy measurements had been made in the marginal ice zone. These measurements were typically single hydrophone measurements that did not address depth dependence or directionality. Some noise measurements at shallow and deep depths had been made that showed lower level low frequency noise near the surface because of Lloyd's mirror surface image interference effects. Whether from ice-camp measurements or from drifting buoys, which had hydrophones typically at 100-ft depth, most of the noise measurements were averages over given times at selected frequencies rather than full bandwidth time series. Measurements over the years, primarily from deep basin buoys, had been accumulated by Mr. Beau Buck, and had provided a set of standard empirical curves representing the 5th, 50th, and 95th percentile noise levels.

These curves were the standard against which all noise measurements were compared.

Early propagation loss measurements were made primarily using SUS charges at spring ice camps. Most of the results were reported as one-third octave band averages of data recorded on 100-ft-deep receivers. It was thought that there was little or no depth dependence to transmission loss above frequencies affected by surface image interference. Due to the complications involved in putting SUS charges through the ice, data from long tracks with dense spacings were difficult to acquire.

A modeling system was not in place prior to the AEAS Arctic Program efforts. Propagation models were not designed to account for ice scattering loss, and the proper way to model ice scattering loss was not known. Arctic environmental databases were incomplete and unvalidated, and not in a form to be easily used by the propagation models.

Early systems performance efforts relied on average monthly noise levels and empirical models to describe propagation characteristics. These were not unreasonable for areas from which the data had been taken to develop the empirical models, but were inappropriate to address such issues as shallow-water propagation and depth dependence. No data were available to address the effects of signal and noise fluctuations, noise directionality, and signal coherence on performance.

AEAS Arctic Contributions

Extensive details of the AEAS Arctic contributions are described in a document generated at the completion of the AEAS Arctic Program, and only a summary of the highlights is presented here.

Field Operations and Instrumentation Highlights

The AEAS Arctic Program conducted, managed, and participated in major data collection exercises in the Central Arctic basins; in the Chukchi, Lincoln, and Beaufort Seas; and in the Atlantic and Barents marginal ice zones. Designed to fill data voids and to address critical system-related issues, the exercises yielded a large and valuable collection of high-quality environmental acoustic data sets that covered a variety of high-priority conditions, including deep and shallow water, areas with bottom slopes and ridges, pack and marginal ice zones, and areas with rough and smooth ice covers.

Due to the extremely harsh conditions and the long distances involved, major efforts were required in planning the experiments and coordinating the logistics to ensure success. Nine major ice camps for AEAS data collection were established in the Arctic. Marginal ice zone exercises used the most varied platforms, including P-3 air operations (1985–1992), submarines (1986, 1988–1992), and small ice camps that operated in conjunction with icebreaker support (MIZEX 88, CEAREX/SIZEX 89). AEAS experiment locations, periods, source and array details, and summaries of collected and processed data are provided in Table 1.

Many technical advances were made during the program in an effort to improve the quality, quantity, and applicability of the acquired data. These advances included new acoustic sources and source calibration techniques, more robust acoustic experiment plans, and improved recording equipment for manned camps, winter-over systems, and air operations.

Acoustic recording equipment that had been used in AEAS open-ocean experiments was upgraded and modified for Arctic use. The long tracks of unbroken propagation measurements and ambient noise time series needed for understanding Arctic environmental acoustics necessitated long-duration, battery-powered recording systems that could be deployed to remote ice camps or left unattended for long periods. Rugged, low power environmental and acoustic data-recording equipment was eventually designed, fabricated, and successfully used in ice camps. The recorders accommodated many channels for recording multi-element arrays and had large dynamic ranges to accommodate the unpredictable variations in ambient noise.

The expanded video cassette recorder (EVCR) was originally developed under AEAS sponsorship to serve as part of the AMIGO (Acoustic Measurements Instrumentation for General Operations) airborne data collection system. At the heart of the EVCR system were commercially available, consumer-grade recording electronics with bit-error rate specifications that far exceeded existing analog recording standards, at much lower prices. Furthermore, the VHS tape-recording medium was inexpensive, smaller, and readily available worldwide. The EVCR units were powered by 12 Vdc, ideal for low-noise recording during the manned ice camps. Variations of EVCR systems included the standard airborne system with its 18 channels and 2-kHz bandwidth, a 100-channel unit with 350-Hz bandwidth per channel, and other combinations between the systems. During the later years of the program, the digital data recorded on the systems could be played back on an inexpensive audio processor.

While EVCRs provided low-power, inexpensive digital recorders for manned ice camps and air operations, the AEAS Program also realized the need for long-term (entire ice season) recordings of ambient acoustic noise in the Arctic. This need was met using recording systems based on the Geotech analog tape recorder, modified to record samples of ambient noise over a full year. Later systems recorded 26 channels at 0.1 inch per second (ips). Systems were developed that could be buried in a floating ice floe or moored to the bottom under the ice cover and recovered a year later. *These systems supplied an unprecedented observation of ambient noise in all seasons, thus facilitating one of the objectives of the AEAS Arctic Program: filling in the seasonal picture of Arctic environmental acoustics.*

The AEAS Program devoted considerable effort to planning and to the field operations to assure that the necessary environmental measurements were made to support the acoustic data being taken. Ice roughness data from submarine digital ice profilers were collected by the Arctic Submarine Laboratory. Meteorological measurements conducted at ice camps included self-recording weather stations (wind speed and direction, barometric pressure, temperature, and solar radiance); these measurements were also required for aircraft operations. Oceanographic measurements taken included current speed and direction (S-4 current meter) and conductivity–temperature versus depth). Shallow-water camps included fathometers for water depth and grab samples to obtain information about the seafloor. Another large-scale asset utilized in both planning and post-analysis was data from satellite imagery. These data provided information on ice cover and drift, as well as lead orientation and statistics.[43]

In summary, over the years, the AEAS Program developed innovative approaches to Arctic logistics, acoustic sources, and recording systems. With each experiment, investigators learned lessons and applied them to future data collections, thereby providing the data needed to understand the effects of the unique Arctic environment on ASW systems performance.

---

[43] The National Oceanic and Atmospheric Administration has a polar weather buoy program that transmits weather data to receiving stations from buoys deployed in various regions of the Arctic. Although no acoustic data are available from these buoys, they do provide a considerable resource that has been used in interpreting long-term acoustic ambient noise data.

Propagation Data Highlights

The AEAS program successfully collected low frequency acoustic propagation from numerous central Arctic and Marginal Ice Zone (MIZ) locations bounding the extremes of the Arctic environment in terms of ice roughness and water mass types. This program was the first to collect continuously sampled Arctic transmission loss with range and concurrent under-ice roughness profiles.

Table 1. AEAS Arctic Data Set Summary[44]

Year/ Station	Location	**Data Collected**								**Receiver Information**	
		SUS	TL	AN	SVP	Ice	Bathy	Met	Geo	Array Type	# Phones
1985/ RUBY	Canadian Basin	•	•	•	•	•	•			VLA	5
		•	•	•						HLA	26
RUBY WO	Canadian Basin		•	•						VLA	5
				•						HLA	19
1986/ TANYA	Eurasian Basin		•	•	•	•	•			VLA	18

---

[44] In the "Data Collected" columns, *SVP* is sound velocity profile, *Bathy* is bathymetry, and *Geo* is geology. Other terms were defined previously.

			•	•						HLA	50
			•	•	•		•			Circular	16
1986/ RED	Chukchi Sea		•	•	•	•	•	•		VLA	12
			•	•						HLA	13
1986/ ROSE	Chukchi Sea	•	•	•	•	•	•	•	•	VLA	33
1988/ JERRY	Lincoln Sea	•	•	•	•	•	•	•	•	VLA	36
			•	•						HLA/ SBL	31
			•	•						HLA/LBL	13
1988/ ELLEN	Lincoln Sea	•		•				•		VLA	48
1988/ MIZ	Greenland Sea		•		•	•	•	•		HLA	15
										VLA	18

1989/ CEAREX/ SIZEX	Barents Sea			•						VLA	7
				•						HLA	7
	Greenland Sea			•						VLA	7
				•						HLA	7
1989/SHIRLEY	Lincoln Sea	•	•		•	•	•	•	•	VLA	32
89-90/ Winter-over	Lincoln Sea			•				•		VLA	17
1992/ GATOR	Beaufort Sea		•	•	•	•	•	•		VLA	17
			•	•						Geophones	4
			•	•						Accelerometer	1
			•	•						Sonobuoys	Many

85-92/ ICEX/ MIZEX	Denmark Straits		•	•						Sono buoys	Many
	Greenland/		•	•						Sono buoys	Many
	Central Arctic	•	•	•						Geo/ Sono buoy	Many

Propagation data were collected from arrays of sensors either suspended from the ice surface or mounted on the seafloor. The arrays provided the capability to determine horizontal and vertical directivity and, for the first time, had the spatial resolution to accurately measure depth dependence in these Arctic environments. Data showed depth dependence similar to model calculations. These data have been studied to better understand Arctic propagation with emphasis on Transmission loss depth, range, and frequency dependence.

AEAS data have provided the basis for improving prediction capabilities at low frequencies by enhancing/updating/correcting environmental databases used in these predictions, and by providing the basis to modify the surface loss function to more accurately estimate losses from surface scattering.

Ambient Noise Highlights

The AEAS ambient noise data analysis presents a coherent picture of the complicated role the convolution of ice noise sources and Arctic propagation characteristics plays in determining the ambient noise spectral, temporal, spatial, and directivity characteristics. Ice cracking generates most of the ambient noise, and the size of the cracks and their speed of propagation determine the frequency characteristics. A major event, such as pressure ridging, will have cracks of varying sizes and propagation speeds. The noise source has a radiation pattern consistent with a dipole main lobe pointed toward the seafloor. The energy then propagates in the Arctic environment. The observed ambient noise is the sum of all the energy propagated from all the noise sources.

The spectral shape of long-term averaged ambient noise generally decays, as frequency squared. Long-term time averages of horizontal directionality in the deep basins tend to be isotropic unless there is a persistent source of ambient noise such as a distant active shear zone or storm front. The spectral and directivity characteristics of short-term averages may indicate the type of noise sources, as well as their distance from the observation point. Nearby noise sources reach the receiver via direct paths and provide a flatter, or "whiter," spectrum and higher vertical arrival angles than distant sources. More distant noise sources reach the receiver via optimized propagation paths, which tend to be the deepest cycling paths supported by the Arctic waveguide.

The vertical arrival structure of noise dominated by these more distant sources varies with the depth of the receiving array. Vertical directionality from near surface (typically ice mounted) arrays shows a small characteristic local minimum at 0° (directly horizontal) and peak energy arrivals at the maximum angle supported by the waveguide. For arrays near the bottom, the peak is at the horizontal, since the dominant ray paths are turning at this depth. Although there is substantial vertical noise directionality, Arctic ambient noise levels are fairly depth independent, except for Lloyd mirror surface image interference effects.

The short-term temporal variability of the Arctic ambient noise was perhaps the biggest surprise of the AEAS data analysis. The ambient noise observations cover a 50-dB dynamic range, sometimes within the span of a single day. There have been many observations with ambient noise changes of 15 dB over an interval of a few hours. This variability dominates sonar performance in the Arctic. The median Arctic noise levels are consistent with open-ocean sea state 0 levels.

Ambient noise tends to increase with the passage of storm systems. The more frequent occurrence of storms in the Arctic fall and winter helps to account for the seasonal nature of the ambient noise. The ice conditions are also seasonal, more brittle in the winter and more plastic in the summer, and this affects how the ice fails under stress.

Ambient noise tends to be the lowest in the summer, increase in the fall, remain high in the winter, and exhibit the greatest variability in the spring. Over the 10–1000 Hz band analyzed by AEAS investigators, ambient noise showed a strong interfrequency correlation and a strong spatial correlation for frequencies below 100 Hz.

In addition to the effect of ice cracking and sound propagation on noise characteristics, another set of ambient noise mechanisms is superimposed in the marginal ice zone. Here, the noise peaks at the ice edge, and the more concentrated the ice edge, the more distinctive the peak. Eddies and biologics are commonly found in abundance in the marginal ice zone. The noise is more variable in the presence of eddies, and biological activity generates another unique set of noise signatures.

Propagation Modeling Highlights

Beginning in 1984, the AEAS program's Arctic modeling efforts focused on developing and validating a standardized prediction capability for the Arctic that included Arctic-specific databases. AEAS had the task of providing a modeling capability that was accurate and easy for Fleet operators to use anywhere in the Arctic, as well as sufficiently detailed for R&D design analyses. The result of these efforts was ICECAP (ICE CAPable) for passive prediction modeling and ICE CAUAMP (ICE Capable AUAMP), renamed ASPM (Acoustic System Performance Model), for active prediction modeling. To bring these capabilities to Fleet users, AEAS sponsored their approval by establishing the Oceanographic and Atmospheric Master Library, which configuration managed Fleet models and databases. The ICECAP ice loss model was accepted by the Master Library as the first and only Arctic Navy standard. ICECAP performed exceptionally well during ICEX exercises and was successfully used to allow R&D developers to optimize system designs.

Many of the AEAS data sets provided transmission loss as a function of range and depth. Agreement between the data and ICECAP predictions for many of these data sets was outstanding, with the models accurately predicting the depth structure and the fluctuation rates. This was especially true for the deep-water sites where propagation paths with very shallow grazing angles are not dominant.

Good model predictions require an accurate description of the environment. In lieu of *in situ* data, the ICECAP databases provided a good baseline. However, for low frequencies in shallow-water environments (low enough that bottom interacting energy dominates the propagation), the Arctic is no different from any other ocean and requires a very detailed geoacoustic description of the bottom. The oceanography of fronts and along shelves is often too complicated for the sound speed database to handle well. For propagation through such oceanographically variable environments, *in situ* measurements are required.

Systems Performance Assessment Highlights

The AEAS Program provided the capability and a methodology to assess the performance of both current and future operational systems anywhere in the Arctic. Specific performance estimates have been provided for air operations in the MIZ and in the pack ice; for SSN operations in the Lincoln, Chukchi, Barents, and East Greenland Seas; and for surveillance systems in the deep Central Arctic and the shallow Lincoln Sea. The approach had been to take the propagation measurements and use them to determine ice scattering loss and to validate acoustic propagation models. Then noise measurements were combined with the propagation measurements (or propagation models and databases) and systems characteristics to estimate and improve performance for submarine, air, and surveillance systems.

While experience in the open ocean was often applicable to SSN and surveillance issues, the air community faced a more severe problem in getting traditional sensors into the water under the pack ice. The best solution to this problem turned out to be the geobuoy, which consisted of a vertical velocity sensing geophone mounted inside a standard A size canister with a 12-inch brass spike for imbedding in the ice to couple the geophone to the ice cover. During ICEX operations in 1986 and 1988–1990, severe problems were associated with the geobuoy's performance. These problems caused the buoy transmission to be received only when the aircraft was within extremely close range of the buoy and caused the buoy to be acoustically overdriven a large part of the time.

Analysis of the AMIGO-collected geobuoy data was instrumental in quantifying the problems and in the development of improvements to both the rf transmission and the acoustic sensitivity. Geobuoys used after 1988 included improvements to the rf transmission. Those after 1990 had reduced acoustic sensitivity, allowing much better determination of the absolute received levels, as well as improved detection performance.

Experiments were designed and the data from them analyzed to quantify the difference in performance between sensors in the water column (sonobuoys) and sensors in the ice (geobuoys). These comparisons showed a performance advantage for the sonobuoy that increased with increasing frequency. However, at the relatively low frequencies of most interest in the Arctic, arrays of geobuoys were used to achieve enough array gain to make up for that advantage. The development of the propagation model ICERAY allowed the determination of the most important paths for sensors in the ice and provided a means to predict their performance.

Conclusion

The AEAS Program sponsored the development of equipment, the collection of data, and the analysis of data to provide the Navy guidance on how to operate in the Arctic and the modeling capability to assess the performance of its platforms in that hostile environment.

CHAPTER Twelve
Data Analysis and Equipment

General Comments

The equipment used by LRAPP can be divided into two separate categories:

- Equipment developed without LRAPP funds (but used by LRAPP and improvements paid for by LRAPP).
- Equipment developed with LRAPP funds.

Some of the equipment was only for making scientific measurements; other equipment was focused on trying new techniques that might be transferable to the military. The measurements, of course, were focused on the terms of the sonar equation. LRAPP had to measure transmission loss, sound speed as a function of depth, ambient noise (both omnidirectional and directional, also as a function of depth) and bottom loss. Some of the information was used as inputs to the different models also developed under LRAPP, and some was used for comparison with the model predictions.

Data analysis is closely linked to equipment. While the skills are focused in different areas of expertise, they are so closely aligned that it is beneficial to consider them together. The data analysis community looks on the experiment definition, preparation, and actual performance with considerable interest, but without much activity. However, when the experiment is complete, and the massive amounts of data are then handed from the "experimentalists" to the

> The director of ARL:UT throughout much of the LRAPP years was Dr. Loyd Hampton. He maintained that the integration of personnel with backgrounds in basic research (6.1), exploratory development (6.2), and advanced development (6.3) made a team that was fundamentally well balanced and could quickly address the LRAPP problems and provide innovative and successful solutions. He was proven right.

data analyzers, their work begins in earnest. These efforts require that the data be organized, reduced, reorganized, and then finally analyzed. This work was done primarily under the direction of ARL:UT. Although many organizations provided support in special areas, the one organization primarily responsible to LRAPP for that function was ARL:UT.

Data Analysis

In the beginning of the LRAPP experience, the data analysis function was spread out among many organizations. Substantial progress was made only when Dr. Gaul recognized that data analysis, as well as all of its associated functions, demanded the same level of preparation and coordination that the at-sea exercises required. The key point was in recognizing that the experimentalists and the data analysts had to work together, and that was the only way in which success could be assured.

The data analysis process was performed by ARL:UT, whose many members contributed to the process. They had a long history of supporting the U.S. Navy in the area of underwater acoustics with a focus on high-frequency sonars. They were cajoled into supporting LRAPP by Dr. Gaul, and once they entered, proved to be a major mainstay in many areas, but with particular emphasis on data analysis and all its attendant functions. Perhaps an equally important role played by ARL:UT was in the comparison of the final data sets to the model outputs from the modeling community. ARL:UT had skills in the experimental arena, but their capability in the modeling area was especially notable.

The experimentalists had always developed their equipment and put their data in a format that was convenient for them. This practice caused a situation where many measurements obtained similar data type, but were done by different organizations and archived in different formats. Any organization attempting to integrate the data into a consistent format

> Mr. Jack Shooter of ARL:UT would usually become upset with the data received from the experimentalists. He always assured LRAPP that it was probably impossible to reduce the data, and even if it were possible, it would take a very long time. He always was able to read the data and reduce it, and do so in a very short period of time. But he stayed in a perpetual state of anxiety.

had major difficulties. Further, most of the data were “raw.” In this sense, “raw” means not reduced into standard data formats.

This situation became a standard problem that ARL:UT always solved, but only with considerable effort. Finally, the problem became so serious that the experimentalists and the data analysts were called in and forced to resolve their differences. These differences were only in how the data were formatted and recorded.

Once the experimentalists made some modest but real changes to meet the requirements of the analyzers, and once computer programs were developed that made the raw data capable of being reformatted so that it could be rapidly reduced and then analyzed, the problem disappeared. Not only did the problem disappear, but also the time required for data reduction and analysis became substantially shorter. Although it never became trivial to organize, reduce, reorganize, and then to analyze massive amounts of data from an at-sea experiments with many different participating organizations, the close interaction between the experimentalists and the data analyzers made the process almost pedestrian.

How I got to Meet a
A Big-Company Vice President

When we (ARL:UT) first joined LRAPP, a well-known company had recently inherited the ACODAC systems, transitioned from Woods Hole. The ACODAC was a deep-water remote recording system developed by Woods Hole for recording ocean ambient noise and acoustic sources to measure long-range propagation.

The ACODACs were the main recording systems for the very large field exercises sponsored by LRAPP, and the recorded data represented millions of dollars in terms of fleet assets (ships, aircraft, planning, etc.).

The basic plan called for the company to deploy the ACODACs, record the data, and send a copy tape to us at ARL:UT for data processing. That all sounds very reasonable, protecting the high value master tape—worth millions. The only snag was that the first copy tape sent to us was blank! This caused consternation in almost every way possible. We were accused of being obvious incompetents, not able to play back a simple tape recording or, at best, we had inferior misaligned hardware incapable of playing back such a precision instrument.

The next thing I knew I found myself on the way to Dallas (the North Plant) with our "blank" tape in tow, so that the company's engineers could demonstrate the quality of the copy recording. I arrived late at night and was introduced to the company vice president who was to supervise the demonstration. I think I was supposed to be impressed, but I was too dumb to know what such a person represented—the power to open and close production lines employing tens of thousands. Well, naturally, we played back the blank tape, and I never saw the VP again. This incident plus a few other recording problems eventually led to a special contract with a different company in Denver to do nothing but COPY TAPES!

Jack Shooter

Equipment[45]

The Floating Instrument Platform (FLIP) was the primary data collection platform used during PARKA. FLIP was a very unusual ship. It was very long, and relatively narrow. When it got to its destination, it filled its ballast tanks, and the bottom part of the ship sunk. FLIP then was in a vertical position. In effect, it was an enormous vertical buoy, looking like a long stick, floating vertically in the ocean. This unique ability allowed FLIP to make measurements in very large storms, with very little movement. The biggest danger to FLIP was long merchant ships which would come over to see what was this very unusual object floating in the ocean. Some times they would come so close that there was real danger of a collision.

The ACODAC system[46] was not used during PARKA but came later. The ACODACs were not only large (on the order of 36-inch diameter pressure vessels) but also contained very large analog tape recorders. The tape recorders and their associated electronics could operate for a period of 30 days, according to a schedule that could be programmed. They were multichannel recorders.

The ACODACs were attached to a string of hydrophones that had been placed in the water column. The hydrophones were located at many different places in the vertical. The signals received by the hydrophones were recorded and analyzed after the experiment was over. The systems were continually developed, and scientists used more advanced technology as it became available.

VEDABS (Versatile Experimental Data Acquisition Buoy System)

One of the extensions of ACODAC was VEDABS, the primary acoustic measurement system used by the ASW Environmental Acoustic Support (AEAS) Program from 1982 to about 1993. VEDABS was developed by engineers[47] at the former Naval Ocean Research and Development Activity

---

[45] The author is indebted to Dr. Darrell Milburn, Dr. Norm Gholson, and Mr. Ken Dial for many of the details on the equipment.

[46] The ACODAC was developed by Dr. Scott Daubin of Woods Hole.

[47] Dr. Darryl Milburn, Dr. Norman Gholson, and others.

(NORDA),[48] Stennis Space Center, Mississippi. Table 2 lists the AEAS measurement exercises conducted between 1982 and 1991. The number of systems deployed in each of these exercises is also shown. *It is said with some substantial degree of pride and professionalism that there never was a failure of the equipment during that entire period.*

VEDABS, a self-recording system, was used to make long-term ambient noise measurements and acoustic propagation measurements. It contained a 28-channel analog tape recorder that could record up to 153 hours. The tape recorder and associated control electronics and batteries were housed in a spherical-shaped, aluminum pressure vessel referred to as an internal pressure vessel (IPV). The system could be deployed as a vertical, a horizontal, or an "L" shaped array.

---

[48] NORDA is now a field activity of NRL. It was also briefly known as the Naval Oceanographic and Atmospheric Research Laboratory before it became part of NRL. For clarity and consistency, however, this facility will be referred to as NORDA throughout this document.

Table 2. VEDABS Deployments

(1982–1991)	
Exercise Name	**Systems Deployed**
OUTPOST RESOLUTION (Apr/May 1982)	4
OUTPOST CREOLE (Jul/Aug 1982)	4
OUTPOST GRANDE (Mar/Nov 1983)	3
OUTPOST REPRISE I (Oct/Nov 1983)	2
OUTPOST REPRISE II (May/Jun 1984)	4
NORTH PACIFIC EXERCISES (Aug/Sep 1984, Apr/Jun 1985, Sep/Nov 1985, Apr/May 1986. Nov/Dec 1986)	10
BROADBAND '85 (Jul 1985)	2
NORTH ATLANTIC EXERCISE (Sep/Oct 1986)	1
FIXED/FIXED (Sep/Oct 1986)	2
NORPAC '86C (Sep/Oct 1986)	2
NORLANT '86 (Sep/Oct 1986)	1
NORPAC '87A (Sep/Dec 1986)	1
NORLANT '87 (Mar/Apr 1987)	2
BROADBAND '87 (May/Jun 1987)	5
NORPAC '88 (Feb 1988)	3
OUTPOST SUNRISE '89 (Sep/Oct 1988)	3
LFA5 (Jul 1990)	2
NATIVE I (Aug 1990)	2
NORPAC '91 (Winter 1990)	2
CST5/LFA7 (Jun/Jul 1991)	3
Total Number of Deployments	58

Figure 1 shows the vertical and horizontal VEDABS mooring configurations. The deployment of the bottom-mounted (horizontal array) system is also shown in the figure. The vertical mooring system was deployed using the "anchor-last" technique and is much easier and more reliable to deploy than the bottom-mounted system. Figure 2 shows a typical bottom mounted system deployment. Figure 3 shows the typical vertical VEDABS array configuration and also the primary parts that comprise both mooring systems. Some photographs of VEDABS taken during several of the AEAS measurement exercises are shown in Appendix C.

> **An Exciting Engineering Test**
>
> My first at-sea experiment was a VEDABS engineering test scheduled out of Bermuda. After a couple of days checking out equipment and calibrating, we were ready to go. Over the next three weeks, at least three hurricanes came through the area. When the weather was clear, the *Bartlett* had engine problems. When the ship was considered ready, the weather wouldn't cooperate. After almost a month, we finally came home with good VEDABS data.

The VEDABS array is constructed from Kevlar electromechanical cable and H-78A hydrophones. The electromechanical cable contained twisted pairs of 24-AWG wire that was used to hard wire the hydrophones to the IPV. The hydrophones[49] have a sensitivity of –182 dB (re 1 V/μPa) over a frequency range of 10 Hz to 10 kHz. Each hydrophone consists of cylindrical ceramic elements, preamplifier electronics, and current drive circuitry. Each hydrophone uses one pair of wires for both power and analog signal.

ADABS

The AEAS digital acquisition buoy system (ADABS) was a 128-channel system that incorporated a digital tape recorder. The ADABS system used the same hydrophones, array, and IPV as the VEDABS. The main difference

---

[49] Developed by NRL's Underwater Sound Reference Division in Florida.

between VEDABS and ADABS was the signal-conditioning electronics, A/D conversion, and tape recorder. A more detailed comparison of these two systems is given in the Table 3.

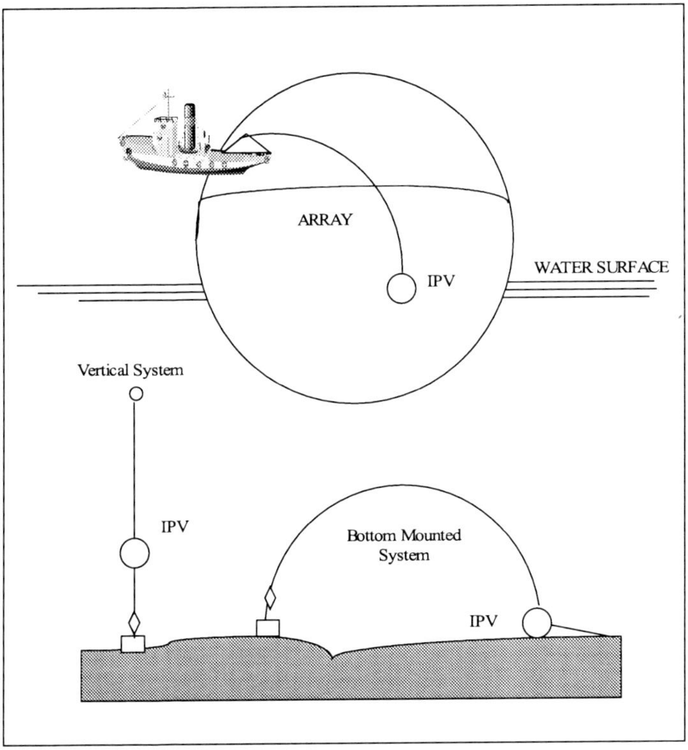

Figure 1: Horizontal and Vertical VEDABS Deployments

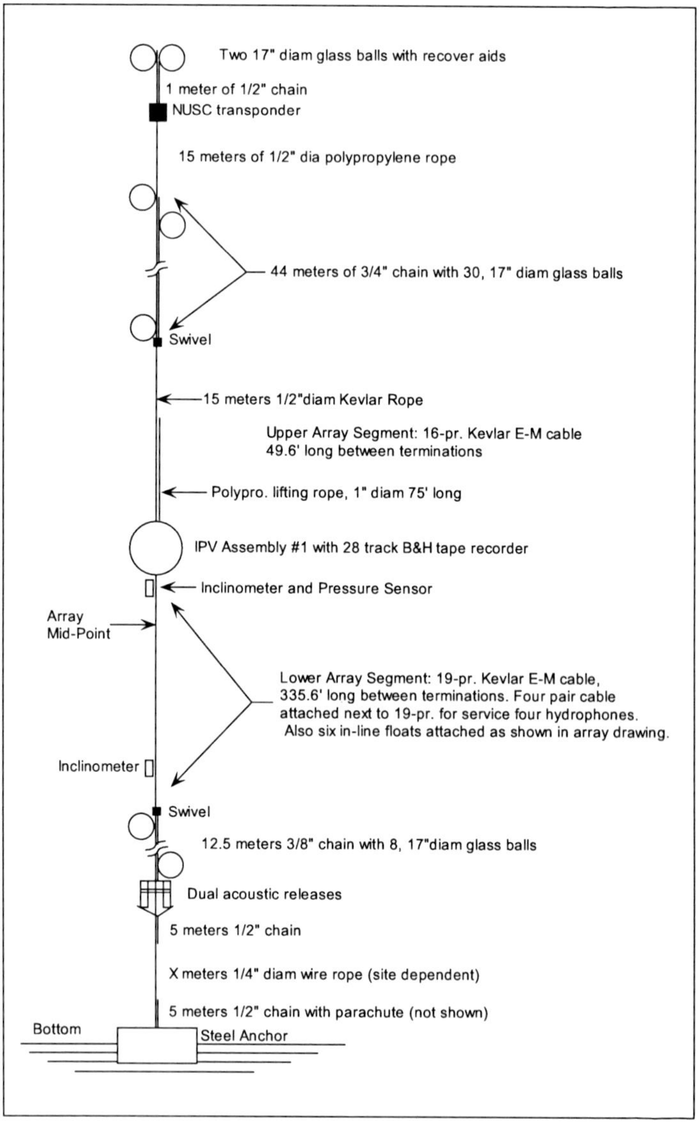

Fig. 2: Bottom-mounted system deployment.

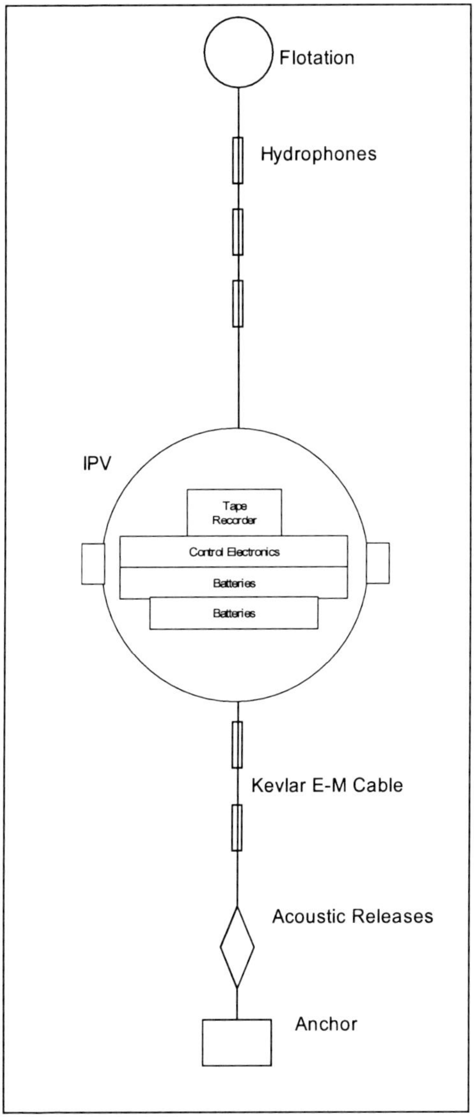

Fig. 3: Typical VEDABS Configuration.

Table 3. Comparison of system specifications.

	VEDABS	ADABS
Recorder mode	direct	digital
Acoustic bandwidth	5 Hz to 1 kHz	5 Hz to 0.5 kHz
Number of acoustic channels	26	128
Dynamic range	30-40 dB	90 dB
Internal calibration	yes	yes
Channel-to-channel	1 dB and 5°	0.5 and 5°
Recording time (hours)	153	111

AMIGO Airborne Data Collection

In the mid-80s, there was keen interest in understanding the area around the Juan de Fuca Straits, which are at the North West corner of Washington State on the Pacific Coast, to explain the loss of contacts by the VP squadrons operating there. What was needed was a way to measure the propagation and noise over an area with such differences in bathymetry and shipping. Making those measurements from an aircraft seemed the ideal way, but the regular Fleet P-3s at that time had an automatic gain control in their recordings and no way to measure absolute sound pressure levels.

The AEAS Program sponsored the development at ARL:UT of the AMIGO (Acoustic Measurements Instrumentation for General Operations), a roll-on, roll-off system certified for flying on P-3s. The original system used an analog instrumentation recorder, but soon transitioned to the expanded VCR, which made use of off-the-shelf audio processors and video cassette recorders.[50] This lighter weight system, with its improved dynamic range and use of inexpensive tapes, was the forerunner of the systems used in the Arctic as well. The OUTPOST ENCORE exercises in the Juan de Fuca area provided the propagation and noise measurements needed to characterize

---

[50] Developed by Dr. Norm Gholson.

that area and explain differences in performance in different sections of the area useful for both the submarine and air communities. The system continued to be used for many other experiments: Marcot; Broadband 87; ICEXes 85, 86, 88, 90, 91, and 92; Glover; and Critical Sea Test 7.

AEAS Arctic Measurement Systems

NORDA's Ocean Engineering Branch was also funded by AEAS to design, develop, and deploy a number of acoustic measurement systems in two of their Arctic measurement exercises: one in the fall of 1984 (ICEX 85), the other in the spring of 1986 (Outpost Heritage 86). The rugged systems used at the manned ice camps were based on the Expanded VCR technology that Dr. Gholson had developed for the AMIGO system, and it accommodated up to 100 channels of acoustic data from horizontal and vertical arrays. The acoustic array used in these measurement systems were constructed of Kevlar electromechanical cable and H-78A hydrophones. These systems were used at Ice Camps Ruby, Tanya, Red, Rose, Jerry, Ellen, Shirley, and Gator over the years from 1985 to 1992. They were also used for marginal ice zone (MIZ) data collections during MIZEX 88 and CEAREX/SIZEX 89.

For the "winter-over" systems left at Ruby in 1985 and in the Lincoln Sea in 1989, modifications were made to the Geotech analog recorders. They were slowed to increase the amount of data that could be recorded and set to cycle on and off to enable samples of data to be collected throughout the year.

Towed Arrays

Towed arrays became the trademark of the LRAPP efforts, and several towed arrays were used by LRAPP during its existence. The ITASS arrays and the TASSRAP arrays, however, were built before LRAPP was formed. They were built in ONR and the Navy laboratories. The man who had the vision was Dr. Marvin Lasky, who is probably entitled to be designated as the "Father of Towed Arrays" in the U.S, Navy. Some arrays continued to be built by the Navy within the system commands, and DARPA built an array called the Large Acoustic Marine Basic Data Array (LAMBDA).

In the Navy the long towed arrays attached to a surface ship made the ships look as if they had tails. Thus, towed arrays in the navy were known as tails.

Towed arrays were built because of the requirement to eliminate the noise from arriving at the sensor from all directions. While the mathematics can become formidable, in essence, the concept is simple. Arrays can be considered to be a series of sensors, all identical, arrayed along a line. The distance between the elements depends upon the frequency of interest. If the outputs of each sensor is added to the others but with a slight but specific time delay, then the signals will be enhanced from one direction and degraded from another direction. This is the concept of electrical steering. Therefore, it is possible to construct a piece of equipment that takes all the outputs of all the sensors in parallel, and for each different branch, make the delays different. The result is a single physical array of hydrophones that look simultaneously in many different directions.

The device that performs this magic is called a beamformer. The original beamformers were physical devices with large cylinders and magnetic delay lines that had connection points to the signal processing over its entire surface. However, as electronics became more sophisticated some complex and sophisticated beamformers were based entirely upon electronics.

One problem still remained, however: the distance between the elements[51] was still fixed for a particular frequency. However, someone came up with the idea of "nesting" arrays. In other words, build arrays for three frequencies (or as many as desired, but three was generally the number used). The idea was to construct directional sensor systems with broadband omnidirectional devices as their elements. To achieve this, the omnidirectional elements were placed in a line with fixed, but different, intervals. Sensors with different characteristics and different sensitivities could be achieved through the selection of the appropriate elements. Some elements were used for one array only; some elements were used for two arrays and, finally, at least one—the center element—was used as an omnidirectional element for all the arrays. This configuration solved the problem of using different frequencies.

Now, towed arrays were conceptually able to look simultaneously in many directions at once (and with computer controls these directions were variable and could be selected at will) and were designed for different frequencies. There remained the issue of constructing the electronics necessary to perform this magic. However, that was accomplished, along with the use of very narrow filters to look at some frequencies of particular operational importance, as well as developing the ability to store the data.

---

[51] In underwater acoustics, the "elements" were omnidirectional hydrophones.

They Said It Couldn't Be Done

LRAPP was responsible for demonstrating that towed arrays, particularly the long towed arrays, operating from small seaworthy ships could acoustically detect and hold submarines over great distances for long periods of time. *LRAPP not only demonstrated that it could be done, LRAPP did it, and did so repeatedly in different oceans in the world, in different seasons, in conjunction with naval ASW forces.*

To achieve this stunning success, LRAPP had to develop the equipment, all the ancillary support equipment (data recorders, beamformers, and the like), along with the models and operational tactics that in slightly modified form was delivered to the Fleet for regular daily use by the operational

The original data collection devices were analog tape recorders. These devices were extended and developed until digital recorders capable of recording enormous quantities of data arrived. The development of these devices came from LRAPP requirements, as well as other programs where recording data for long periods of time was required.

LAMBDA

The original LAMBDA was developed and built by DARPA. LRAPP used it in one or two experiments. The man who was most responsible for the development of the original LAMBDA and its successors was Dr. Henry Aurand of NOSC. Dr Aurand was one of the original workers with Dr. Hersey at the inception of LRAPP. It was clear that in order to find Soviet submarines it would be necessary to have very narrow beams (approximately 1° or less). This need for such narrow beams required the towed arrays to become of remarkable length. The original ITASS arrays were approximately 600 feet long; the final LAMBDA array was over 3 miles long!

Some additional problems were associated with towed arrays that required considerable skill and imagination to overcome. Towing an array that exceeded 3 miles in length (that included not only the array itself, but also the ancillary vibration isolation modules, power cables, etc.) was not difficult for the ships that were used. These ships were oil-rig resupply boats, commonly called "mud boats," which are extremely seaworthy vessels used in the seismic array industry; they were developed, supported, and relied upon by the off-shore oil industry. In fact, long towed arrays were used in the offshore oil industry before they were used in the U.S. Navy. The Navy insisted that the offshore oil industry was ahead of them technologically, although the offshore oil industry always came to the Navy to see what advances they had made. Both communities were making advances almost simultaneously, but had different objectives.

The difficulty in towing long arrays is that the towing ships must proceed slowly. The average speed of advance (SOA) was about 3 knots, or about 3 miles per hour. Prosecuting such a slow SOA was entirely foreign and to a U.S. Navy combatant vessel commander, so the Navy was pleased to use civilian ships to perform this mission.

Another technical problem also challenged the LRAPP community: these long towed arrays wiggled. As they were pulled through the water at different depths, the arrays began to develop bends. They were waves in the array, caused by its own motion. These wave motions were side-to-side and up-and-down. The beamformers were built on the assumption that the arrays were straight and horizontal. This difference between the theoretical array attitude and its real one would cause errors in the look directions and the beamwidths, and furthermore would become time dependent. Such a problem could be easily solved if the positions of each sensor was known in three-dimensional space as a function of time.

It turned out that the problem was relatively easy to solve, more or less. Perfection is always the goal, but is never achieved. If very short bursts of acoustic energy from a noise source could be received by every sensor, then the time delays in the signal arrival would allow the determination of all the sensors. The method used was to fire mortars from the ship, lobbing mortar shells into the ocean, having them explode at some known depth, and then after repeating this in a short time (a few seconds apart) with three or more mortar shells, the positions of each array sensor could be determined. These positions would be entered into the computations, and the wiggles could then be taken into account.

Towed-array-generated beams were so narrow that all noise could be mathematically rejected from every direction except in the direction of the beam itself. So, if a submarine was the only object in the beam being considered, it was possible to detect the submarine, even at great distances. In fact, the beams were so narrow that we were able to "look between the ships," where the ships were the surface merchant vessels that were the primary generators of the noise detected by the arrays in that particular frequency band. The other part of that problem was that the submarine was in the beam only a relatively short time, so, it was necessary to look at many (hundreds) beams very rapidly and very frequently. Computers solved that problem.

Straighten the Tail

The LAMBDA array was impressive: 15,000-foot-long active elements, vibration isolation modules (VIMs) forward and aft on the array, and very heavy 2.25-inch, torque-balanced tow cable. The front of the array was very heavy. The arrays were towed at 1.5 knots, and operated at a depth of over 4000 feet. Based on the heading sensors in the array and the "mortar technique," it was determined that the array was not straight when under tow. I was asked to try to straighten it out.

The tow cable acted like a rigid rod. Ship-induced motion transferred directly into the array. This was verified with accelerometers on the transom of the ship and in the forward VIM section. There was a one-for-one correlation in motion. This was because the tow cable took a 90-degree turn on the bottom. The ship's vertical motions were translated into horizontal motions into the array. The problem was how to keep the tow cable vertical.

I felt that if we could place a heavy enough weight on the bottom of the tow cable, the tow cable-VIM connection would be come a knuckle in the system. I contacted Morris Akers at NOSC and asked him to find a heavy, ball-like mass that weighed about 1500 to 2000 pounds. We conducted tests with this mass and it worked. Shipping-induced motion was damped out in the VIMs. Rigging the wrecking ball was difficult, especially out at sea. Therefore, I accepted a compromise approach: lead collars on the bottom of the tow cable that distributed 2000 pounds of weight over about 100 feet in length. This configuration could be pulled up through the sheave, along the conveyor tray, through the level wind onto the winch without any assistance.

Another problem addressed was array undulations. One concept considered to address the problem was to place resistive force, a.k.a. drag, on the array end section. A single large sea anchor would work, but past experience had proven that they tended to kite outward until they "spill," then kite in another direction. An alternate approach, which we used in the end, was to place six sea anchors, 6 feet in diameter, in series. They helped to straighten the array.

Sea trials were to be conducted off Santa Cruz Island in deep water. The test was scheduled at 6:00 a.m. on Saturday. That early time created a personal problem—my son was graduating with honors from high school in Texas, and I didn't want to miss the event. I had missed enough birthdays, holidays, and family events for LRAPP. I advised Dr. Gaul of my problem, and his reply was, "You have to be there. This is YOUR test." I went to the graduation and family celebrations, took a 1:00 a.m. flight from Houston to Los Angeles, picked up a rental car, and drove to Oxnard. I had made arrangements for a fast Sport Fisherman to take me to the ship, about 50 miles offshore. I arrived at 5:00 a.m., and tests commenced at 6:00 a.m. That was the LRAPP way.

Claude P. Brancart

A Rude Awakening

The Church Stroke I exercise began as scheduled in the summer of 1977. The second day out of port, about 0700, Sam Marshall (NRL) and I (then of NOSC) were abruptly awakened by the deafening sound of an explosion. It seemed to penetrate the stateroom we shared from all directions equally. The magnitude of the noise it made indicated that the source of the explosion had to be nearby. We nearly broke the stateroom door and each other in our efforts to escape.

Our first thoughts were to look for the source of the explosion and to help assist in whatever efforts were needed to save fellow researchers, salvage the experiment, save the ship, or man the life rafts, whichever was the most appropriate for this tragic situation. There was no evidence in the passageway of either the cause or the effects of the explosion.

Sam and I rapidly made our way to the main deck, which was immediately above our stateroom, and unexpectedly found Bob Martin (NUWC) calm and composed. He was standing beside a three-foot metal tube that was supported by two metal legs to form a tripod. An open end of the tube was pointed up into the air at about a 45-degree angle in the broadside direction away from the ship. The other end of the tube was terminated by a small metal plate, which rested on the steel top deck of the ship. The location of this peculiar contraption was at that part of the top deck that formed the ceiling for the stateroom which Sam and I had so rapidly exited only moments before.

To our astonishment, Bob did not share our alarm or excitement. In fact, he seemed to be enjoying the spectacle of two "seasoned" LRAPP seagoing veterans running around acting as if the sky was falling. Bob seemed to savor this spectacle for what seemed, to us, an unnecessarily long time. Finally, with a sinister smile Bob explained that he and his strange contraption were responsible for the explosion that rudely forced us from our stateroom. He explained that his part of this LRAPP experiment was to conduct studies of the effects of array straightness on the performance of the line array that was being towed by the ship they were on. Furthermore, he explained, for these studies he needed an explosive sound source underwater at a range of about a mile from the array along the broadside direction. He further related that his strange contraption was actually a combat mortar borrowed from the Army. The explosion that we had heard on the other side of our stateroom ceiling was the first firing of a mortar round specially modified to explode at a prescribed depth, after being launched from the deck of the ship and arriving at its targeted range.

Bob then suggested to Sam and me that we might prefer to join him in his future firings, all of which were scheduled for 0700 most of the remainder of the cruise. Since there was no hope for us ever to sleep past 0700, we reluctantly accepted the invitation, and Bob welcomed the early morning company that he had been secretly hoping for. Bob, obviously very pleased with himself, described this as a "win-win situation." The disgruntled victims,

Other Equipment

Other pieces of equipment were used throughout the LRAPP at-sea exercises; they were standard Navy equipment: SUS charges, bathythermographs, sound velocimeters, and others. SUS charges were used for the transmission loss measurements and were dropped from P3 aircraft flying specific tracks. The charges could also be dropped from surface ships moving along designated tracks. SUS charges were the source of the signal in the water, and the ACODAC, VEDABS, etc., would be the receiving equipment.

By relating the time of arrival of the signals to the time the SUS charge was dropped and then taking the difference in the energies, the transmission loss from the source to the receiver could be measured. Although the signal arrival strength was carefully measured, it was necessary to assume the signal strength of the SUS explosion, which was a function of depth and type of SUS charge. No one had made serious measurements of this source strength until LRAPP realized that it was crucial. LRAPP made some careful measurements and provided the Navy with an upgrade to their information on the SUS charges.[52]

---

[52] This was performed under the direction of Dr. Marvin Weinstein of Underwater Systems, Inc,

CHAPTER Thirteen
Computer Models

General Comments

The development and use of computer models was one of the major areas of LRAPP focus; the other was at-sea exercises. The two worked together in a symbiotic relationship. In general, both could—and do—exist without each other, but when the two are meshed together in a common focus, their individual value increases dramatically.

Models are attractive to the scientific community for several reasons. First, they are relatively easy to build and are also inexpensive. Many scientists and engineers can build a computer model. For that reason, many technical people are either upgrading an existing computer model, or have built their own computer model. This practice has its good and bad points. On the one hand, it is good because of the substantial interaction between people with similar models. By interacting, they can compare their models; on the other hand, people tend to want to continue to use their own models and denigrate models developed by others.

Models are also attractive to the operational navy for two major reasons: system design and performance prediction. In the area of underwater acoustics these take on an added role since the ocean has a unique effect on sound. The first reason that models are so important is that they assist with the design of equipment. When the SQS-26 was designed, the total effect of the ocean was not appreciated and a world wide and expensive effort needed to be undertaken to develop a database on the ocean bottom since many of the sonar signal waveforms were distorted by the bottom in ways the sonar design engineers did not understand. A world wide sound speed database also needed to be developed since the sonar's detection ranges varied from season to season and region-to-region. The second reason models that models are so important is that the navy cannot measure every ocean parameter that is needed to predict sonar systems performance. Therefore sonar performance models are needed.

If there is a major concern of the operational world anywhere in the technical community and not just limited to naval applications, then it is configuration control. If several people or organizations have ostensibly the same computer model and if they use the same database to serve as the input data, then the results should be identical. However, this is almost never the case. People and organizations have a long history of taking models, their own as well as those of other people, and making small "upgrades." Further, they don't really like the database that is provided, so they make some small changes. Finally, all of these "identical" models are run on different machines! The end result is small and sometimes large differences in the outputs. This is a classic case of anarchy. From an operational point of view, if the same scenario is run by different organizations, and the results are slightly (or more) different, then which model output is correct? Which one should be used? The solution to this problem is all-important.

In the days before mass production (we will use the classic example of musket production), each musket was made according to the same design, but each component was slightly different. The parts were not interchangeable. Introduction of mass production made certain that every part of the same type was identical with its clone, and they were totally interchangeable. The production procedure instilled rigid configuration control. Every musket was identical to every other musket manufactured by one manufacturer. Later, the configuration control process caused the muskets made by different manufacturers to be identical, as well.

This state of control is mandatory for industrial production of physical parts. It is also necessary for computer models when they have reached the operational state, too. That, however, was not LRAPP's role. Rather, it was the development of the models in the R&D sense that was part of the LRAPP function.

Computer models have two major tests that they must undergo: one is verification, the other is validation. These tests are both complex and sophisticated processes in the computer model world; in concept, they are both fairly simple. Verification is the process where the actual internal operations of the models are considered. Does the data flow from one part of the model according to the rules set by the developer? Is each individual component of the model doing what it is supposed to do? While these questions appear to be simple, in practice it becomes difficult with larger models. In a model that is small, such internal data flow is easy to ascertain and check; in models that may number 500,000 lines of code, such a process is almost impossible. Once a model is verified (that is, it is actually calculating what it should, in the steps as required), it can go on to the next process: validation.

Validation is the process whereby the model outputs, or predictions, given a specific set of input data, are compared to "truth." The issue of "truth" is almost philosophical, but in this particular area, it is defined to be measured data.

Let's consider an easy example. Consider a large, horizontal, flat surface (an asphalt parking lot will do nicely) where an experimentalist has a device that will throw a ball (a baseball), so that it always leaves the device at a known angle with the horizontal and a known speed. The angle can be measured with some known degree of precision (with a protractor) and the speed of the ball also is measured (using a radar gun) with some known degree of accuracy. The experimentalist then runs the experiment many times. He notes that the ball travels in an arc and lands at a particular position away from the device. He then measures the distance from the device to the place where the ball first hits the pavement. He is generating a set of data where the measured result is the distance and the inputs for each experiment are the angle and speed.

Now, a modeler will generate (easily) a simple model that will predict the range as a function of the input angle and speed. He also runs the model many times. In fact, he repeats the entire set of experiments.

The process of validation is when the predicted results from the modeler are compared to the measured results from the experimentalist, and they agree to some measure. In this particularly simple example, the agreement will be extremely good. Of course, it will be important to the measurement accuracy that no wind was blowing and that there was nothing really strange about the ball (e.g., golf balls dimpled to help control the boundary layer on the surface of the ball, or if a balloon filled with water deforms in flight, the results would probably not be as good as with a golf ball or a smooth baseball). The point to understand here is that the predicted model results will be compared to real data, and, if the differences between the predicted ranges and the measured ranges are acceptable to the eventual user of the model, then the model is said to be validated.

It does not require much imagination to see that this case is simple and that almost all the variables are under our control. Pick a smooth ball; if it is windy, wait for another day to perform the experiments. Make sure the radar gun works, and that your protractor is reasonably accurate. But when the experiment is run at-sea over ranges of a few dozens to thousands of miles, it is not possible to wait for another day. Further, myriad conditions are not under the control of the experimentalist, and even worse, are possibly unknown to him. Nevertheless, the experimentalist must take as much care as possible in making their measurements, as well as knowing which data will be required for input to the acoustic models.

> LRAPP never forgot that its eventual product would be models that could be used by the fleet.

Several different types of models were developed under LRAPP sponsorship, as well as used by LRAPP in the design of their experiments. Further, these models would be run by major facilities, such as FNWC, and now, even on board ship and aircraft during actual operations. This perception meant that there were conditions that were placed on the models. For example, in the beginning of LRAPP, RP70, a transmission loss model, worked well, but it was painfully slow. In fact, it was a research model and could not meet the time requirements for delivery of predictions to the operating fleets. FACT met those time constraints, but it was designed for use in an environment where there was only one sound velocity profile and the bottom was horizontal. This particular condition is (more or less) met in large areas of the Northeast Pacific (the area of the CHURCH ANCHOR experiment) if propagation was east and west, i.e., at a relatively constant latitude. If propagation was north and south, however, the sound speed profile changed with latitude. Thus, the actual environmental conditions for north–south propagation (more or less) violated the assumptions built into the FACT model. It was clear that the FACT model was not correctly dealing with the actual environmental conditions. Therefore, it was also clear that the predictions of transmission loss would be different from the actual measurements. Inevitably, the question was always, is the FACT model good enough? Only comparison to measured data would provide that answer.

The acoustic models built and used by the LRAPP community essentially tried to predict the values of the terms of the sonar equation, transmission loss and ambient noise. Transmission loss is based upon the geometric positions of the source and the receiver, and the environment. The environment consisted of the sound velocity field and the losses and spreading of acoustic energy from the ocean surface and bottom.

In the development of the transmission loss models, substantial effort was put into the development of databases, especially bottom characterization. The models developed for that particular problem were compared to the SQS-26 bottom-bounce sonar database generated for the NAVOCEANO by Alpine Geophysical. This was a world-wide effort begun long before LRAPP was established. The data were taken at 3.5 kHz (the frequency at which the SQS-26 operates). Since the LRAPP frequencies of interest were in the general area of 25–300 Hz, some remarkable steps were taken to work with the large database.

One of the problems with the 3.5Khz database was that when the data was used at lower frequencies the results became suspect since in some regions of the world. The calculations showed gains of energy and not loss of energy as the sonar signal bounced off the bottom. This did not happen when exact experiments were done at these lower frequencies. A major accomplishment of the LRAPP/AEAS program was the development of a geophysical database called the Bottom Loss UpGrade (BLUG), which is now referred, thankfully, as the Low Frequency Bottom Loss (LFBL) database. This database developed by ARL/UT, WHOI, SAIC and NRL allowed energy to penetrate the bottom sediment at these low frequencies at angles below critical. The energy then returns to the water column. The critical angle concept and the thickness of the sediment added to the database allowed the modeling results to match the at-sea data. It also allowed the database to be used by transmission loss models such as the Parabolic Equation and Normal Mode models that were range dependent.

The calculation of ambient noise is almost the same as transmission loss. It is necessary to calculate transmission loss as part of the calculation of ambient noise, but in this particular case, all the sources are merchant ships, and they are located on the sea surface. However, they are not distributed uniformly over that surface. Approximately 40,000 merchant ships are in existence[53] and about 16,000 of them are at sea on any given day. Their locations in each ocean are dependent upon the season of the year. Because approximately 3000 ships are located in the northern Pacific, the calculation of ambient noise anywhere in the northern Pacific must include both the source level and the transmission loss from the merchant ship on the sea surface to the receiver.

It was gratifying when this calculation was done using only the ship distributions measured in CHURCH ANCHOR and the results showed that the measured omnidirectional ambient noise and the predicted ambient noise were different by only 0.5 dB.

Once the transmission loss and the ambient noise models were in hand, the real problem could be considered. For a given body of water and a given environmental scenario (season), where should the receiver (almost always a towed array array for LRAPP but other passive sonars for the navy) be placed to optimize the probability of detecting a submarine? The placement was not only at which depth, but also which direction the array should be aligned.

---

[53] Numbers gathered in 1973 for the Historical Temporal Shipping (HITS) database, a database of merchant ships throughout the oceans of the world. The HITS model was developed by Dr. Louis Solomon and Dr. Allen Barnes of Planning Systems Incorporated. This was a part of all at-sea measurement programs, and was performed with the support of the U.S. Navy using Maritime Patrol assets, generally P3 aircraft. The source levels of the merchant ships were based on data measured and compiled by Dr. Robert Urick and presented in several of his books on underwater acoustics.

The answer to this question required millions of calculations[54] and required many hours of thought and preparation. But this was exactly why LRAPP had been formed. To some extent, the problem is not quite as difficult as implied. It was known (more or less) by LRAPP and the U.S. Navy that acoustic sensors worked much better in certain areas than others. Detections were made in certain areas and hardly ever in others. Why?

In the Mediterranean, for example, in the western end near the Straits of Gibraltar, it was known through the vision and experiments by LCDR McCloskey that the proper place to put a towed array was south of the straits, almost nestled up to the African coast, and look northward. The main beam would (this was the MFA system) see between (!) the merchant ships as they exited and entered the Mediterranean through the straits. The periods

Mobilizing in England

Outpost Creole I & II were back-to-back exercises in the spring and early summer of 1983. Creole I was a VEDABS deployment in the GIUK gap; Creole II was in the Mediterranean, and the last towed array exercise that SEAS conducted. Both phases required mobilization of a ship and all of the logistic issues that surround that activity. Mobilization of Creole I was in Glasgow, Scot-land, and Creole II in Felixstow, England, on the English Channel.

Charlie Stuart, with little explanation, asked me to call the NAVOCEANO shipping department and order us a C-5 to fly a large amount of stuff to England. So I picked up the phone and called John Childress and told him I wanted a C-5.

First there was silence, and then there was John's hysterical laughter. When he could talk, I assured him I wasn't kidding. It took a couple of days for us to come up with the shipping list (dimensions and weight), and it turned out that a C-5 wouldn't do the job. We used three C-141s to transport the material from Charleston (trucked there from Stennis Space Center), Fort Worth, and Dover to Mildenhall.

Ken Dial

---

[54] This number of calculations does not include the number of individual calculations required to generate the transmission loss and ambient noise databases.

where there were no merchant ships allowed the transit of Soviet submarines to be easily detected.

If the towed array was placed in approximately the middle of the strait traffic and looked toward the west through the straits, then the array would be overwhelmed by the noise generated by the merchant ship traffic. So, in this particular example it was known where to put the towed array and what its orientation should be. The depth of the array could be decided by calculating the expected performance against submarines for that particular season. While the calculation was not a trivial thing, it was substantially easier than the general case.

By the 1980's LRAPP, now called AEAS, was in the forefront of a new sonar modeling thrust; the on-board performance system. In the 1970's the advent of small computers e.g., the NOVA 800, allowed models and databases to be taken on-board ships and the computing done on the ship, with the ship's exact sonar parameters and data measured with on-board BT's or XBT's. There was now very little need to radio back to a shore bases facility to get a range prediction. NUWC had developed SIMUS, with their in-house developed range independent model and databases. NOSC (Now SCC-SD) had developed ARPS. The principal developer was Ken Faucher, (with a little help from Ed Chaika). In 1980 Ken became AEAS modeling manager. NOSC had their own model and databases and gave the hardware to NAVOCEANO , who added FACT and their own databases. FNWC was also using FACT and their own database.

AEAS had developed the SPARS system for use in the surveillance community. Now each ship and aircraft at sea could get it's own sonar performance; the major problem was that the answers were different and the Navy wanted a consistent and correct answer. In 1984 the Oceanographer of the Navy (OP-096) established the Ocean and Atmospheric Master Library (OAML) to independently validate and verify models and database that would go into all Navy at-sea systems. All but two of the first twelve models and databases added to OAML were AEAS developed. Computer technology was so rapid that by 1986, FNWC stopped broadcasting acoustic range predictions.

Computer Models and Databases Developed By or Associated with LRAPP

Some of these models were scientific and used for scientific purposes. Some eventually became Fleet Models, i.e., they were specifically built for daily and frequent use at FNWC or used at-sea for performance prediction. Some models were built only to answer Fleet questions. The following incomplete list has examples of all kinds.

Ambient Noise Directional Environmental System (ANDES)

This model replaced DANES.

Acoustic Signal Excess Prediction System (ASEPS)

Provided signal excess performance predictions for specific SOSUS sites and the SURTASS towed array. Predictions were either for user-defined target locations or were contours for ocean-wide geometries. ASEPS integrated the ASTRAL Transmission loss model, the DANES ambient noise model, the NOSC/ PEP signal processing beamforming model, and several innovative environmental data access components for its calculations. Developed at ARL/UT, SAIC and ODSI..

Acoustic Sensor Range Prediction System (ASRAPS)-Fleet Model (FNWC)

ASW Prediction Areas

These were areas throughout the world that had homogeneous acoustic properties. These included water mass properties (a single sound velocity profile), constant water depth, and a single bottom loss characteristic. These areas employed the Basset-Wolff curves, bathymetric charts, and FACT.

ASW Transmission Loss (ASTRAL)

This was an approximate model to calculate transmission loss very fast. Its predecessor was FACT, which could operate only with flat bottoms and with a single sound velocity profile. ASTRAL could deal with variable range bathymetry (bottom depth changes with distance) and multiple sound velocity profiles. It used an averaging technique with range steps of generally 50nm. Its early predecessor, such as RP70 took one hour to calculate the transmission loss as a function of range for a range of 1500nm. ASTRAL could do the same calculation in one (1) CPU second. Furthermore, ASTRAL predictions, when compared to acoustic measurements, was found to be quite accurate. . In 1987 the averaging was replaced with 1 nm spacing and therefore could be used for tactical transmission loss runs as well as long range calculations, It is still part of OAML.

Basset-Wolff Curves

This model was a series of curves giving the bottom loss as a function of bottom type.(Type 1–5 bottom). Served as the inputs for all bottom interaction calculations until BLUG.

Bottom Loss Up Grade (BLUG)

Provided a systematic approach to the bottom interaction calculation. The calculation was developed by AESD in conjunction with ARL/UT, WHOI and NRL. It was based upon physics and input data from the Alpine Geophysical Data Base developed for NAVOCEANO during the SQS-26 bottom measurement program begun in the early 1960s. Its principal author was Mr. Chuck Spofford.

Directional Ambient Noise Environmental System (DANES)

Developed as a major directional ambient noise model. It used ASTRAL as the transmission loss model and HITS as the shipping model. Developed ODSI; the principal author was Mr. Osborne.

Fast Asymptotic Coherent Transmisson (FACT)

Developed by AESD. Recognized that in a flat-bottom single sound velocity environment the ray traces are repeated cycle after cycle; therefore, it was only necessary to calculate the first cycle, and then repeat at the appropriate distance. In fact, due to the simple environment, it was only necessary to calculate the ray fan in the first fourth of the first cycle, and then to use mirror imaging. This model went through many upgrades, including issues associated with caustics, sources on the surface, and many other secondary but real issues. It became the first Navy Standard Model. FACT was amazingly rapid and turned out to be sufficiently accurate for many environments. Principal author was Dr. Ray Cavanagh.

Historical Temporal Shipping (HITS)

Provided the distribution of all the merchant ships and fishing vessels throughout the worlds' oceans. This distribution is for the summer and winter seasons of 1973. Developed by Dr. Louis Solomon and Dr. Allen Barnes of Planning Systems Incorporated under contract to LRAPP. Published in 1975. Version 4.0 is still in OAML.

ICECAP

Developed by the AEAS Program for predicting performance in the Arctic. The ICECAP ice-loss model was verified using data from many Arctic exercises and has been accepted as a Navy Standard model by the Oceanographic and Atmospheric Master Library.

Low Frequency Bottom Loss (LFBL)

This database is now version 11.0 in OAML and has variable resolution areas.

Multiple Profile Program (MPP)

Upgraded version of RP-70

Navy Interim Standard Sonar Model (NISSM)

Calculated ranges for the SQS-26 sonar. Developed by United States Naval System Laboratory (USNSL) that later became Naval Under-water System Center (NUSC) in New London, CT. This model eventually became an upgrade input to SHARPS.

Parabolic Equation Model (PE)

This was a major step forward in acoustic calculations. The propagation of sound is governed by a differential equation that is hyperbolic in character. For a single frequency source, the differential equation becomes elliptic in character. The implications between the two types are enormous. Hyperbolic equations start from a single condition, and radiate away from the source. Elliptic equations must have the boundary conditions specified everywhere on the boundary to obtain a solution. Both types of equations have techniques that are useful, but difficult to handle. Dr. Fred Tappert, Bell Laboratories, developed the parabolic equation, which is a happy mixture of the hyperbolic and elliptic differential equations, but keeps the useful properties of both. There were many different versions of PE, all with different acronyms. . P.E. 5.1 is now the OAML standard.

PHITAR, PHITAR-V Fleet Model (ran at FNWC)

Technique for Fleet use to obtain detection ranges based upon the ASRAP model. Developed by Mr. Chet Wilcox.

Range Prediction (RP-70)

Developed by CAPT Paul Wolff et al. at FNWC. The first variable profile, variable bottom contour model that produced ray traces and eventually transmission loss. Based upon Snell's Law.

Ship Helicopter Acoustic Range Prediction System (SHARPS) Fleet Model

Provided range prediction for active sonar systems. Operated from FNWC. Developed by LT Terry McCloskey.

SPPS

Provided earliest prototype for ASEPS. SPPS provided SOSUS perform-ance predictions using bathymetrically dependent empirical logarithmic extensions to FACT calculations for long-range transmission loss. SPPS was replaced by ASEPS, which used the more appropriate ASTRAL transmission loss model.

Surveillance Range Acoustic Prediction (SURVRAP) Fleet Model

FNWC calculated the ambient noise. This was then convolved with the SOSUS beamformer to provide SOSUS beam noise predictions. The transmission loss was then calculated. The sonar equation was then used to calculate signal excess. The ultimate result was a prediction of detection opportunity.

Submarine Range Acoustic Prediction (SUBRAP) Fleet Model

This model was combination of SHARPS and TASSRAP. It was applied to the systems and sensors used by submarines.

Towed Array Sensor System Range Prediction (TASSRAP) Fleet Model

TASSRAP calculated transmission loss both at sea and at FNWC. It then used the ambient noise predictions and generated towed aray system detection ranges.

Wagstaff Noise Rose

A technique developed by Dr. Ron Wagstaff to take ambient noise as measured by the LAMBDA array and deconvolve the measurements into directional noise.

WIT (Wagstaff Iterative Technique) Algorithm

Systems Performance Predictions

The System for the Prediction of the Acoustic Response of Sensors (SPARS) was initiated at ARL:UT under the AEAS Program as an R&D prototype system to investigate the benefits of providing the surveillance community with a real-time, on-site prediction. The concept proved to be extremely successful and grew in scope as the system was used to support fixed SOSUS arrays and later the mobile array, SURTASS. In Fiscal Year 1986, an agreement on the capability of the SPARS system, which would transition out of an R&D mode to become a Fleet Support system supported by SPAWAR[55], was made. It was to be supported by PMW 180 which was the new name of PME-124. The work, started under AEAS sponsorship,

[55]SPAWAR is a System Command, and is the successor to NAVELEX.

continued to support the IUSS community with predictions to support fixed distributed systems design and performance, ADS mission planning, and low-frequency acoustics. The Advanced Surveillance Acoustic Prediction System (ASAPS) supported the development work required to provide the Fleet with a hybrid acoustic prediction system using synoptic noise and environmental data. This work was also the basis for standard Navy performance predictions for use onboard submarines and surface ships.

CHAPTER Fourteen
Proof in the Field

General Comments

In any program there comes a time when the final product of many people must be tested to see if the primary objective of the program was successfully met. This is "where the rubber meets the road," and all the efforts made over a number of years and after money has been allocated and spent are finally put to the penultimate test: does the system do what it is supposed to do? This time came for LRAPP in the summer of 1980 in the Mediterranean Sea.

> **The Move to Mississippi**
>
> For a variety of reasons, some political, some not; the base of operations for several Navy programs was moved from Washington, D.C., to Mississippi, and LRAPP moved with them. Everybody became residents at the Stennis Space Center, operated by the National Aeronautics and Space Administration, circa 1977.
>
> At that time, the Naval Oceanographic Office was divided into operational and R&D functions. After the move, the R&D group that was formerly part of NAVOCEANO became known as NORDA.
>
> Neither LRAPP's workforce quality nor the outstanding support given to the U.S. Navy were diminished by the move.

The Sine Qua Non of the LRAPP Effort

This effort was being tested in the Mediterranean in the summer of 1980. Rear Admiral (RADM) C. R. Bell, USN, (now VADM Bell, USN Ret.)[56] was then in charge of the part of the Sixth Fleet responsible for U.S. submarines. VADM William Small was Commander, Sixth Fleet (COMSIXFLT). He specifically ordered RADM Bell to know where all the submarines were, all the time. Note that this was not just U.S. submarines,

---

[56] Much of this information is based upon a personal interview with VADM Bell.

but all the submarines of all the nations, both our allies and our adversaries.

The principal concern for the submariners during exercises and operations was water management. Although the Mediterranean Sea is very large compared to a submarine and the chances of one submarine running into another was remote, it was still possible. However, at certain positions (for example, the Straits of Sicily) the waters become relatively narrow and the possibility of an underwater collision, though still not high, substantially increases. VADM Small was right to be concerned, and he had given RADM Bell an order that was quite difficult to implement.

The successful implementation of ASW required integration of all the U.S. Navy's assets. The glue that held all of these assets together and forged them into a successful force was intelligence.

CTF 66 became the center for all ASW coordination for the Sixth Fleet. In general, all platforms were used in the ASW operations, but they did not have common goals, tactics, or guidance in their daily operations. RADM Bell, as the senior flag officer, through his chief of staff, Captain McCloskey, and using a new Joint Operational Control Center in Naples, Italy, forced coordination of all the ASW forces, including surface ship towed arrays from CTF 60. This clear, well-defined operation was remarkably successful.

RADM Bell pointed out that one fights as one trains. We knew that we were superior to the Soviets in both men and platforms. Nevertheless, the Soviets had good submarines. Mid-frequency acoustics (300 Hz) had not been used regularly before this time. This fact contributed to our poor or moderate success in detecting Soviet submarines. At this time there was one new tool added to RADM Bell's operational assets: the Mid Frequency Array (MFA). It was specifically designed to operate at 300 Hz. It was the belief of RADM Bell, and strongly supported by CAPT McCloskey, that when all the assets were used in a coordinated manner it was possible to actually solve the ASW problem in real time, for long periods of time. All the assets available to the U.S. Navy were used: submarines, surface ships, VP aircraft, SOSUS, and the MFA.

The use of the MFA was a major event in the Mediterranean. The professional naval personnel were very happy to have this tool, even if it was run by civilian scientists. Naval communications personnel were on board the towed array ships and were responsible for all communications. The scientists on board the ships operated within the U.S. naval operational constraints and provided excellent information. The physics and the science were correct. Due to the LRAPP experiments and practice with the equipment, it was well known where to put the MFA. The MFA allowed continuous detection and contact of all Soviet submarines throughout each of the Mediterranean basins in which it was deployed.

CHAPTER Fifteen
Reconstitution and Death

General Comments

Things were going well. Although Dr. Gaul had left, his place was taken by several very capable men who followed in his footsteps. The evidence of the ability of towed arrays to detect and track submarines was proven. The fleet had used them as part of their operations in several areas and found that ASW could be accomplished effectively. SOSUS, the towed arrays, and the other branches of the operational fleet could detect, track, classify, and maintain a killing solution for incredibly long periods of time. The battle of ASW between the U.S. Navy and the Soviet Union submarines was essentially decided. Through persistence and the integration of technical and operational assets, the devotion of small numbers of military personnel and civilian scientists and engineers had won the day. Everything worked, and the team could serve as the prototype for standard military ASW operations.

Although relatively fierce political battles were still being fought, this was a standard problem, and everyone involved knew that as long as ASW remained at the top of the priority list, things would never be too bad. ASW had been at or near the top of the priority list for over 20 years and showed no sign of slipping from its position.

SURTASS, which had competed with the MFA with some vigor (since it was felt that it could suffer some programmatic changes), had triumphed. They were in the system for continued development, production, and purchase. The ships were being designed to tow them.

Then the Soviet Union collapsed. From the U.S. Navy's point of view, the Soviet submarines went from being a very serious threat, which was the responsibility of the Navy to counter and control, to being a possible minor annoyance. In a very short period of time, ASW slipped from its high priority to just another problem to deal with. This slippage was reflected in the reorganization of OPNAV. The organization responsible for ASW, OP-095, was renamed and became known as Director for Naval Warfare, and was renumbered to become OP-08. Further changes followed in the next few years.

All the information that had been gathered by LRAPP was still highly relevant and would be used if the nation ever needed those data in the future. But ASW was no longer of great importance. Since LRAPP always claimed that the knowledge of the environment was extremely important and played a major effect of the design, development, and employment of systems, the program continued. Now issues arose that had only been listened to with half an ear. The Navy declared that mine warfare and the Arctic were also very important. LRAPP, as an advanced development program began to focus on those areas, but there was no great interest or support from the Navy.

LRAPP remained in ONR. However, with time and the budget battles within the Navy, changes were made in the programmatic structure. LRAPP had been designated Program Element 63785N since its conception and birth. Now, that number was struck from the list, and LRAPP was funded from another program element as part of a more general task. When a program loses its program element designation, it means that it has died.

It should be said that many of the veterans from the ASW days are still associated with ONR and LRAPP, although it is now called by another name. The same skills are still used and applied by men and women to solve the operational and technical problems encountered by the operating forces. But LRAPP as an entity is dead.

CHAPTER Sixteen
Lessons Learned

General Comments

The Navy has a wonderful system called Lessons Learned. After each exercise, a meeting is held where participants discuss what was learned during the exercise. These "lessons" are then written down and distributed to the relevant parties. The lessons can and do range from safety, to planning, to tactics, to operations, to data, etc. In fact, these lessons are really a compendium of anything that the people involved feel should be passed along and used to make future exercises more successful. It is intended that the only mistakes made in the future are new ones and are not repeats of old mistakes. This process works; and its value has been confirmed in countless examples. In discussions with LRAPP participants it has been possible to distill the Lessons Learned in the course of the LRAPP experience. These lessons are certainly applicable to other future programs.

Two Principal Lessons

Lesson One: Accountability and Fiscal Resources

Delegate to the Program Manager (PM) the necessary authority to implement all the actions required for the program to achieve its objectives. Hold the PM responsible for all aspects of the program. The PM must report directly to his superiors, who hold him personally accountable for the overall progress and eventual success of the program. In addition, the program must have the fiscal resources necessary to accomplish its stated mission objectives.

Lesson Two: Team Construction and Operation

Create a comprehensive team from all participants. This team consists of people and organizations from government, industry, and universities. They all have roles to play, depending upon their expertise, and each and every one is responsible for doing so as part of the team. This procedure applies to all facets of the program: technical, operational, financial, contractual, and managerial. Every team member, regardless of their area of responsibility, is informed about each and every phase and aspect of the program. They all participate in the discussions about the original concept, initial planning meetings, detailed organizational meetings, implementation of the program steps, on through to the final report. This approach, which takes considerable time and effort, will cause all people associated with the team to have the success of the program as their objective. They will understand how their individual specialties integrate with the overall operations. Those people and organizations who refuse to play as an integrated part of the team are removed swiftly, without compunction, by the PM. There is no appeal over his head.

What Specific Detailed Lessons Did We Learn?

Nonsystem Programs

The RDT&E process focuses on specific systems. These systems, regardless of where they originally start their development, eventually migrate to the SYSCOMs, as is appropriate. However, certain programs are not system-specific. The Navy is organized along platform lines (submarines, surface ships, aircraft, etc), not functional warfare missions areas (e.g., ASW).

The only OPNAV organization that was functional in character was OP-095, the Director of Antisubmarine Warfare. This organization cut orthogonally across all platform lines. In the same way certain programs, such as LRAPP, focused on the effects of the environment on system performance. LRAPP served as an advanced development home for all science and technology that was not system-specific but that could be applied to system programs of many types. Such a program that was mission oriented, based upon science and technology, and dependent upon fundamental physics needed to continue and serve as a repository for knowledge that could be applied to multiple systems.

Contracting

Contracts must be made with organizations that have people with skills needed by the program. In R&D where science and technology are the fundamental basis for possible success, it is the skills, artistry, and devotion of individual people that increase the probability of program success. It is specifically not the organization with the lowest price. Further, the contracts must have substantial flexibility that will allow for major changes in direction as the needs arise. Finally, the contracts must be of sufficient length so that once the contracting process is finished, only appropriate monitoring is required. This flexibility will allow the work to proceed with all due deliberate speed without major administrative impediments.

Contractors

Contractors should be integrated into a cohesive and smoothly operating team. This team consists of government and industrial components. They cannot be successful if both sides distrust each other. If contractors cannot operate in a fully open and trusting manner, then they must be terminated. No contractor should have more than 10% of either the operating funds or the responsibility for their portion of the program. Further, once a contracting organization has shown that they are worthy of trust and deliver products that work—on time and within cost—then their positions should be secure.

Military Liaisons

The military, academic, and scientific worlds are not naturally compatible with each other. They operate according to different criteria, different time scales, and have different measures of success, yet they are natural allies in the defense world. To achieve the goals and objectives of any program that wishes to provide products of utility to the armed forces, there must be close and trusting relationships among the participants. This trust can be substantially assured if there is close liaison with military personnel assigned to the scientific program areas, and they must play a significant and active role.

Program Management

Management of the program must be active and pervasive. Any process that selects a major organization to operate as the program manager's representative is doomed to failure. The government officials who are the program managers are in charge, always. This responsibility entails substantial control of the program and almost unceasing meetings and travel. The review process must include all the pertinent players, including all the members of the team, both government and non-government.

Navy Laboratories

The Navy laboratories have remained the keepers of the scientific data. They are the organizations that will retain the historical perspective and information for the entire U.S. Navy. They must be included in the program supporters and must be held to the same high standards for performance and delivery of operational products as the industrial contractors.

Mix of 6.1, 6.2, 6.3

Advanced development programs (6.3) are dependent upon the ideas and concepts of the scientific and technology communities. The thinkers that provide information that is applicable to and supports advanced development programs frequently are part of the basic research (6.1) and exploratory development (6.2) communities. They need to be integrated into the advanced development program and should be considered as major assets that should used as much as appropriate. The mingling of representatives of the different communities invariably generates fresh ideas on how to successfully apply science and technology to operational problems.

Senior Management Involvement

Senior civilian and military management need to be aware of the progress of the program, including successes and failures. They must become the champions and advocates of the program. In the budget battles (a continuing struggle), they are the champions that will save the program and allow it to continue. This involvement process requires briefing civilian and military senior management on a regular and continuing basis.

Marriage of Science and Military

The military has operational problems that are extremely difficult to solve. They are bound to operate under certain rules and as part of a culture that is substantially different from that of civilian life. They mistrust civilians and feel uncomfortable around them. Civilians generally find the military peculiar, with strange habits, odd procedures, and a set of values that are substantially different from those of the civilian world. If military and civilians work together with a common goal, then they will find that a grudging respect for each other begins to occur. Civilians will find that the procedures used by the military are not capricious and arbitrary but are based upon substantial operational experience. Military personnel will find that civilians are the best and perhaps only hope that is available to develop science and technology that will provide answers to their operational problems. The two groups must be required to work together; respect will grow on both sides. Those who refuse to work with each other must be ruthlessly removed from the program.

The Holy Grail

Both civilians and military are devoted to the United States as a nation, and the defense and support of the citizens is paramount to both. The primary objective of the program must be pictured as providing an important and measurable part of the defense of the United States as a whole. Focusing the entire program team on their major mission objectives will provide a sense of urgency and devotion that is characteristic of people in combat. They will work together as a team, trusting and depending on each other. Those who do not contribute or who consider their participation in the program as only another job should be removed from the program. In some sense, the members of the program team should enlist for the duration and believe that they can solve the problem and, further, that solving the problem is in the best interest of the United States of America.

APPENDIX A
Exercise Locations

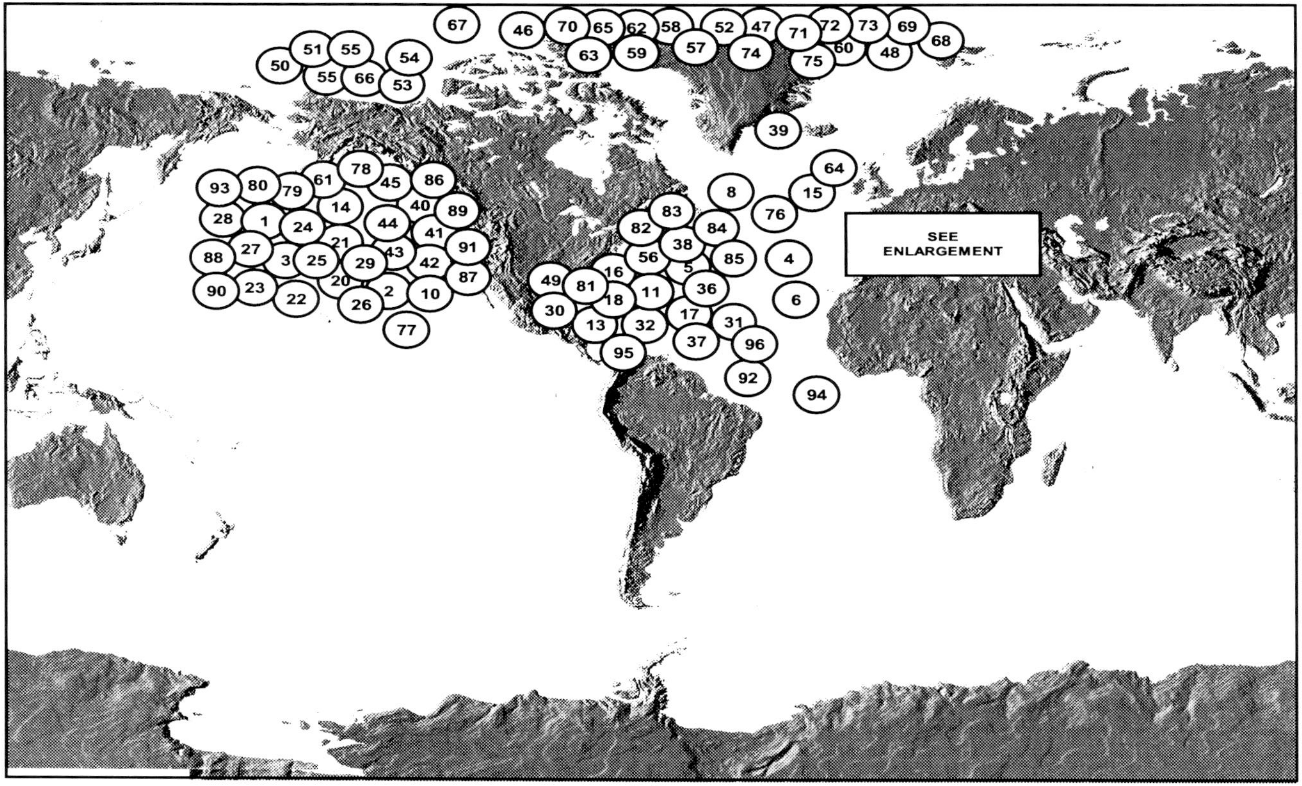
SEE ENLARGEMENT

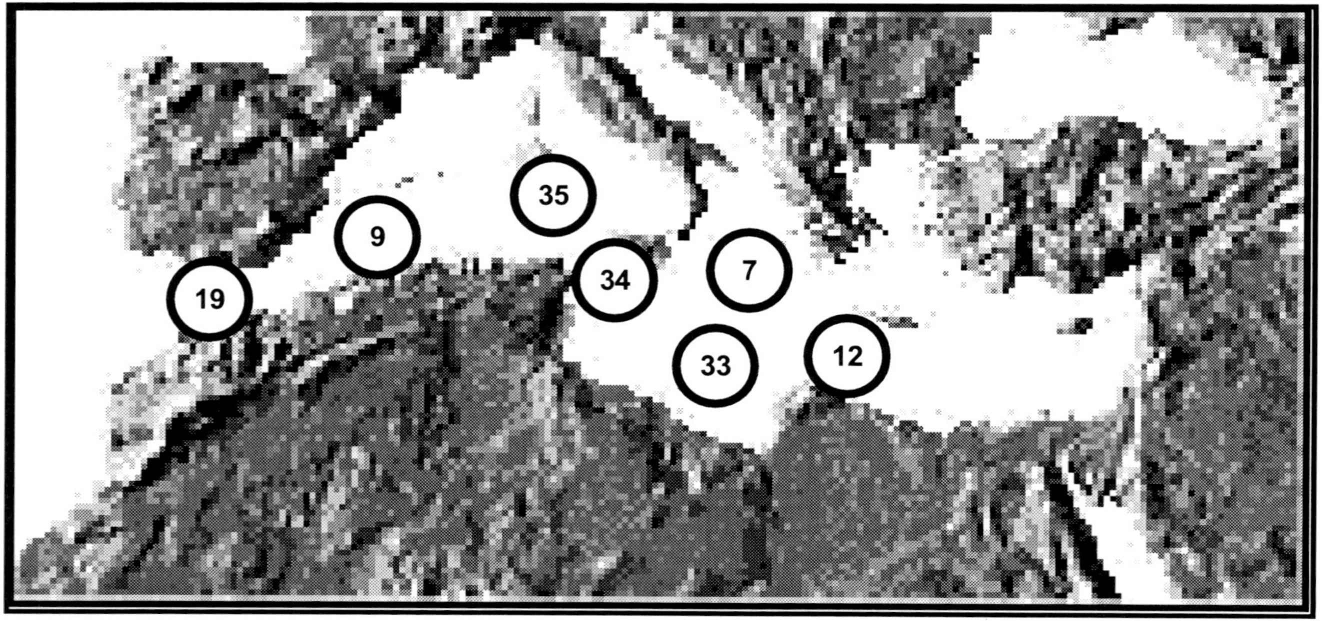

GEOGRAPHIC KEY TO LRAPP EXERCISES

NO.	EXERIMENT	DATE	GEOGRAPHIC AREA	REMARKS	
1	PARKA I	Aug-Sep 68	N of Hawaii	Incl. Sea Spider attempt Rudy Nichols/CS; Gerry Fisher proj coord	
2	PARKA IIA	Sep-Dec 69	NE of Hawaii	Incl. Sea Spider attempt Rudy Nichols/CS; Gene Flesher/OTC	
3	PARKA IIB	Mar 1970	N of Hawaii	Fred Spiess/CS	
4	NEAT I 70?	NE Atlantic	Ralph Goodman		
5	TEST BED	Dec-70	Bermuda	Ray Hasse	
6	NEAT II	71?	NE Atlantic	Ralph Goodman	
7	IOMEDEX	Nov-71	Ionian Basin		
8	NORLANT-72	Jul-Sep 72	E of Labrador	Bob Martin/CS	
9	Alboran Sea	Aug-Sep 72	Alboran Basin		
10	EASTPAC Sea Test		Oct-72	Off San Diego	
11	Bermuda Coherence Test (unnamed)		Oct-72	Bermuda	
12	TASSRAP	fall 72	Med		
13	Church Gabbro	Nov-Dec 72	Caribbean Sea (Cayman Basin)		
14	Church Anchor	Aug-Sep 73	NE of Hawaii	Bud Fadness	
15	Square Deal	Aug-73	NE Atlantic		
16	Blake Test	? 73			
17	MSS Baseline	Oct-73	Virgin Islands		
18	CAPER		Jul-74	Pacific	
19	MASWAP		Aug-74	Med	Terry McCloskey
20	Kent Beacon		? 75	Pacific	Primarily Fleet exercise

21	Church Opal	Sep-Oct 75	N of Hawaii	
22	Fixed-Mobile Exercise	? 76	Pacific	Primarily DARPA
23	CHURCH STROKE I	Jun-Jul 77	NE of Hawaii	
24	CHURCH STROKE II, Cruise 1	Aug-77	Pacific	
25	CHURCH STROKE II, Cruise 2	Sep-77	Pacific	
26	CHURCH STROKE II, Cruise 3	Oct-77	Pacific	
27	CHURCH STROKE II, Cruise 4	Nov-77	Pacific	
28	CHURCH STROKE II, Cruise 5	Nov-Dec 77	Pacific	
29	CHURCH STROKE III (Cruise I)	1979?		
30	GULF OF MEXICO CRUISE	1979		
31	PILOT MEASUREMENTS	1979		
32	RDSS/AVT (Rapidly Deployable Surveillance System)	1979	Caribbean Sea (Cayman Basin)	
33	Big Dipper	Jun-Aug 80	Ionian Basin	
34	SEINE NET I	1980	Strait of Sicily/W. Ionian Basin	
35	SEINE NET II	1980	Strait of Sicily/W. Ionian Basin	
36	WESTLANT 80	Oct-80	NW Atlantic	
37	BERMEX 81	1981	Atlantic	
38	WESTLANT 81	1981	NW Atlantic	
39	OUTPOST CREOLE I	summer 82	NE Atlantic	
40	OUTPOST RESOLUTION	spring 82	West of Washington State Juan De Fuca Ridge (JDF)	USNS DeSteiguer
41	OUTPOST CREOLE II	fall 82	West of JDF	
42	OUTPOST ENCORE	1983	NE Pacific (JDF)	
43	OUTPOST REPRISE I	1983	NE Pacific (JDF)	

44	OUTPOST ENCORE II	1984	NE Pacific (JDF)	
45	OUTPOST REPRISE II	1984	NE Pacific (JDF)	
46	ICEX-85 THRU ICEX-92	1985	Central Arctic & MIZ	
47	OUTPOST AREA 85	1985	N of Greenland	
48	OUTPOST AREA 85	1985	Central Arctic	
49	PROJECT GEMINI	1985	South of Galveston TX	
50	ICEX I-86 Ice camps Red and Rose	1986	Chukchi Sea	Ed Gough. PI
51	ICEX-86	1986		
52	OUTPOST HERITAGE 86 Ice camps Mandy, RubyII, Astrid	1986		
53	OUTPOST HERITAGE	1986	Central Arctic	
54	OUTPOST HERITAGE WEST	1986	Western Arctic	
55	OUTPOST HERITAGE WEST	1986	North of Barrow AK	Dan Ramsdale PI
56	BROADBAND 87 (Broadband Dist. Sys.	1987	Bermuda	Bruce Steinberg PI
57	ICEX-87	1987		
58	ICESHELF-88 (Ice camp Jerry)	1988	Lincoln Sea	Dan Ramsdale, PI
59	ICEX-88	1988		
60	MIZEX-88	1988	Greenland Strait	Paul Bucca, PI
61	CRITICAL SEA TEST II	1989	N. Pacific Gyre	
62	ICESHELF-89	1989	Lincoln Sea	Ice Camp Shirley/Spinnaker
63	ICEX-89	1989		
64	Critical Sea Test IV	1990	E. of Scotland/UK	
65	ICESHELF-90 (Ice camp Panama) (Spinnaker)	1990	Lincoln Sea	Dan Ramsdale PI
66	ICEX 90	1990	North of Barrow AK	

67	ICEX-90	1990	
68	OUTPOST AREA 90	1990	North Pole
69	AREA 91	1991	
	Ice camps Opal, Crystal, Ruby		
70	ICESHELF-91	1991	Lincoln Sea
	Ice camps Fogg and Sudds/Spinnaker		
71	ICEX91	1991	
72	ICEX-92	1992	
73	OUTPOST AREA 92	1992	
	Ice Camp Gator		
74	OUTPOST AREA 93	1993	N of Greenland
75	OUTPOST AREA 94	1994	
76	BLAKE TEST		Atlantic
77	BOTTOM INTERACTION		SE Pacific
78	CHURCH OPAL		NE Pacific
79	CHURCH STROKE I		NE Pacific
80	CHURCH STROKE II		N Pacific
81	CHURCH STROKE III		Caribbean
82	CRITICAL SEA TEST I		
83	CRITICAL SEA TEST III		
84	CRITICAL SEA TEST V		
85	EASTLANT		
86	FIXED FIXED		NE Pacific (JDF)
87	FIXED-FIXED I Exercise		
88	KENT BEACON		NE Pacific
89	MARCOT 2-85		NE Pacific (JDF)

90	NATIVE 90	
91	OUTPOST CREOLE I	West of Juan De Fuca
92	OUTPOST GRANDE	South Atlantic
93	PACIFIC SURVEYS	N Pacific
94	SOUTHLANT	SW Atlantic
95	WESTERN GULF TEST	Gulf of Mexico
96	WESTLANT	NW Atlantic

APPENDIX B
Timelines

Events from 1961-1965 (World, USN, and LRAPP)

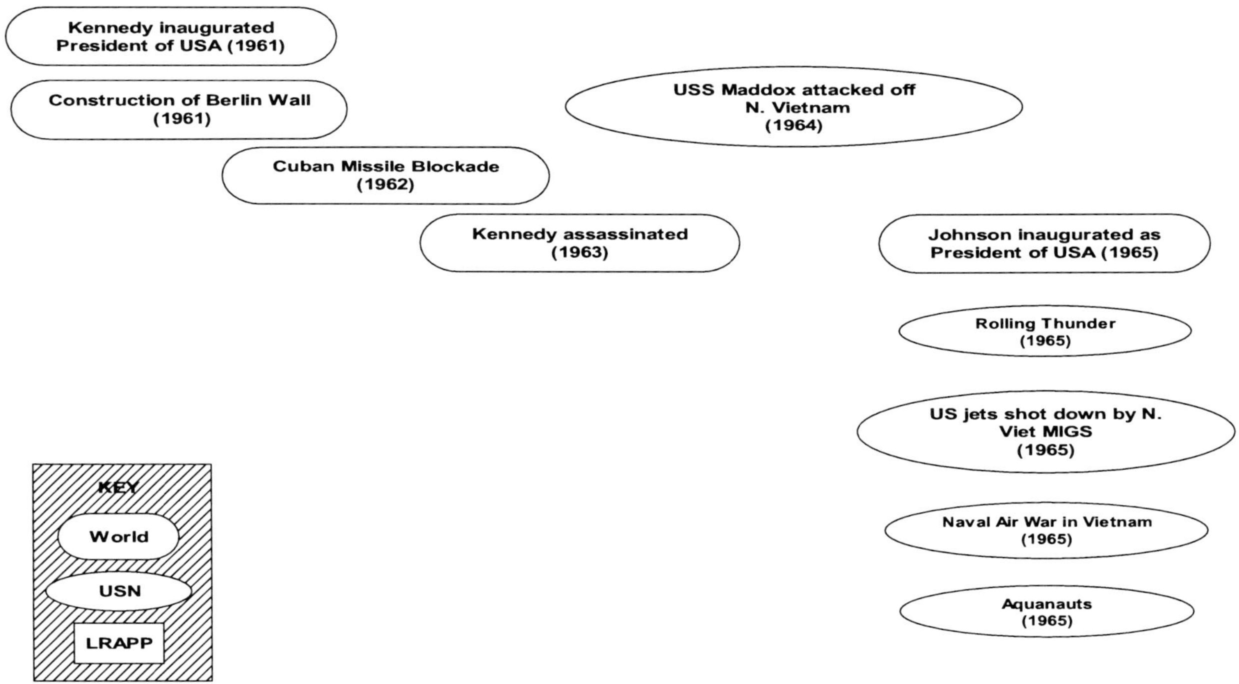

Events from 1966-1970 (World, USN, and LRAPP)

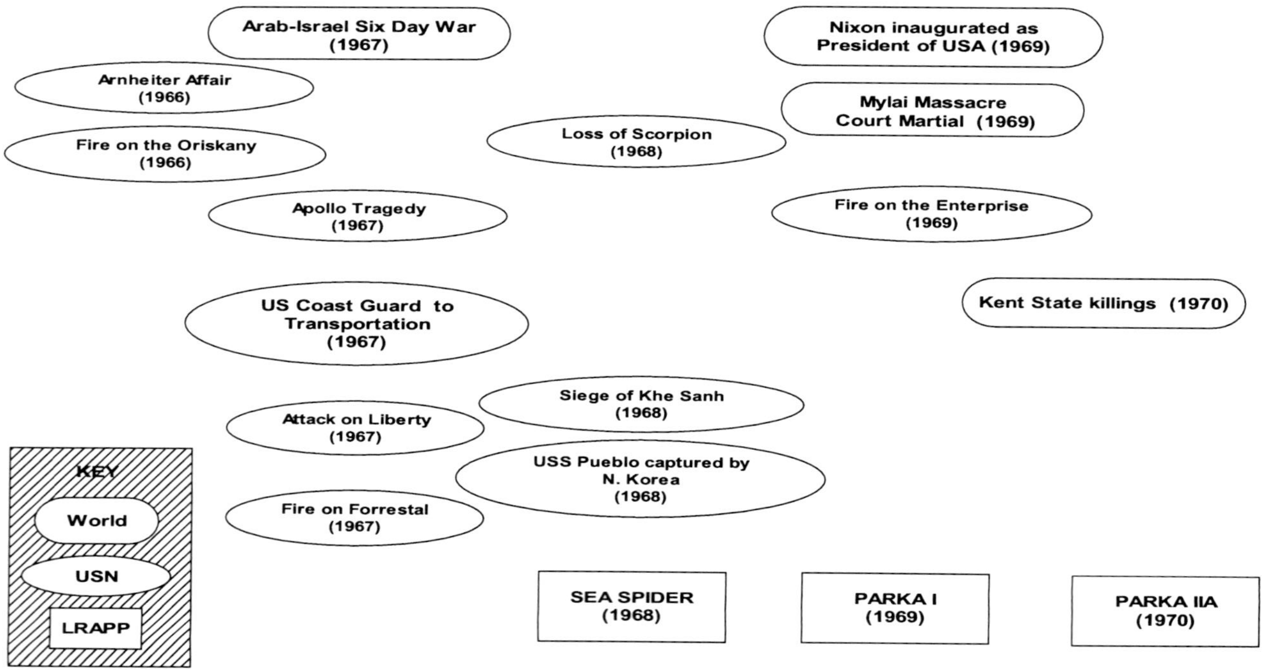

Events from 1971-1975 (World, USN, and LRAPP)

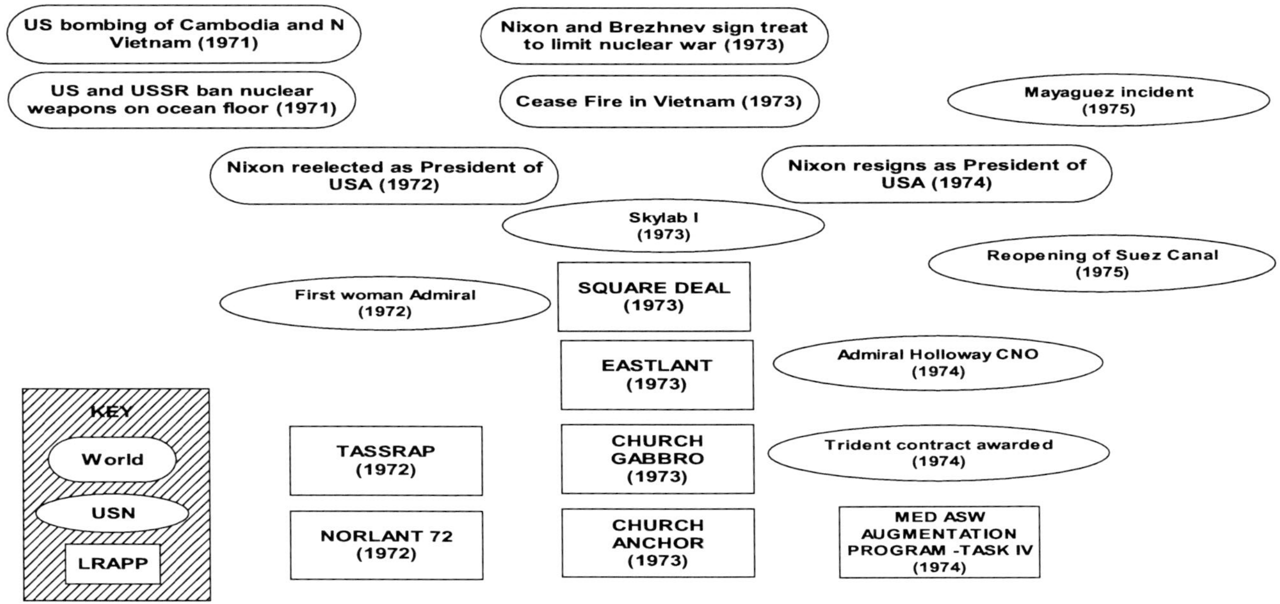

Events from 1976-1980 (World, USN, and LRAPP)

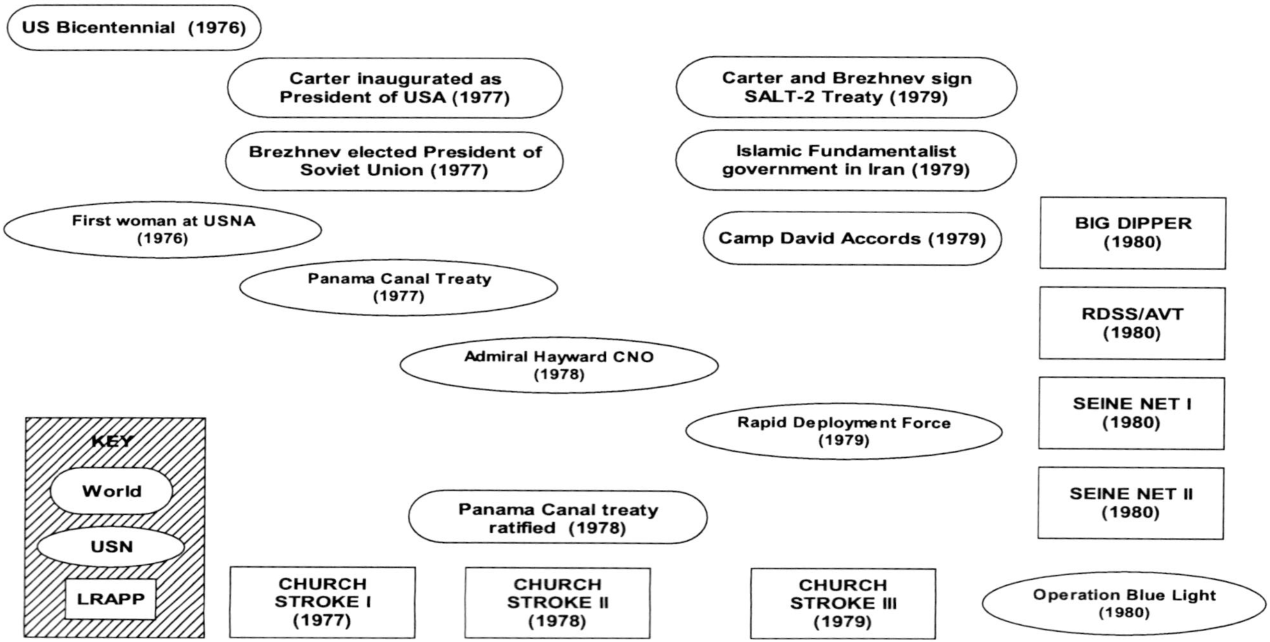

Events from 1981-1985 (World, USN, and LRAPP)

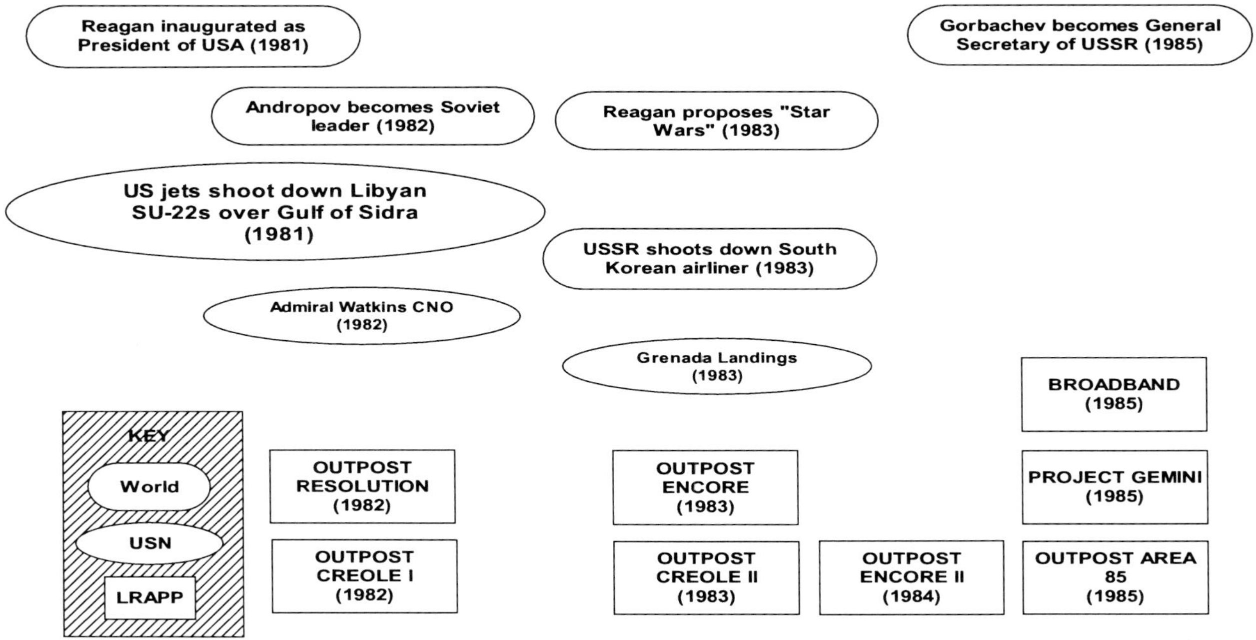

Events from 1986-1990 (World, USN, and LRAPP)

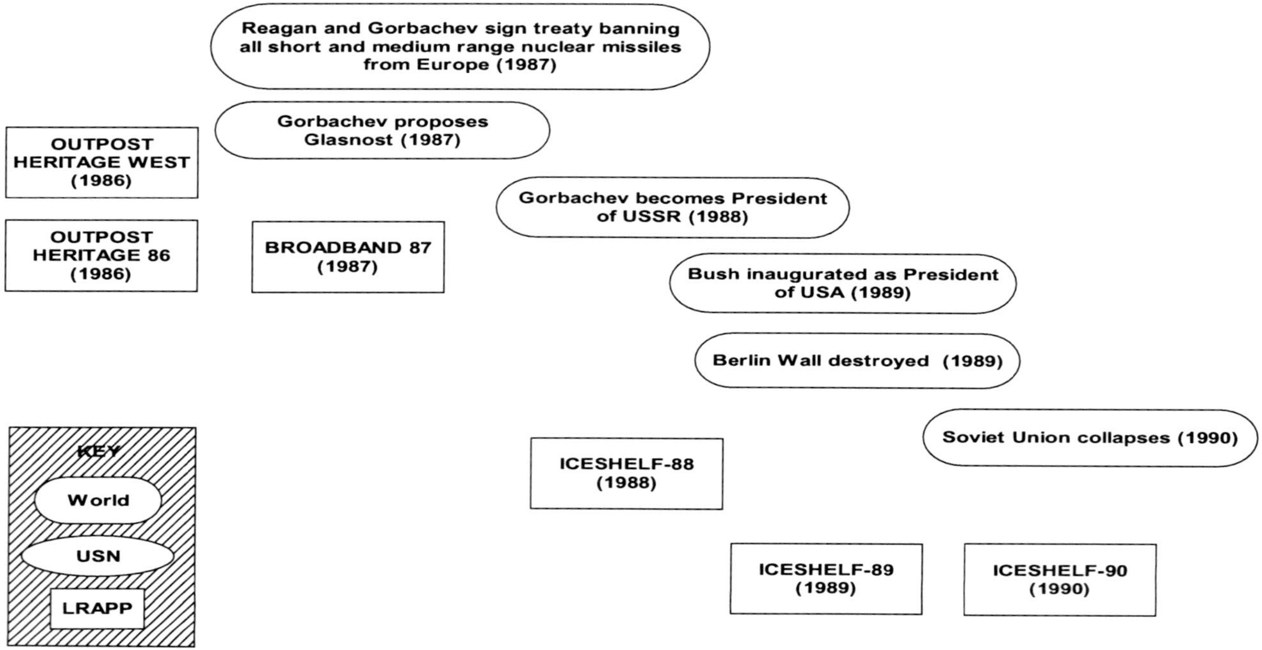

APPENDIX C
Photographs

Table of Photographs

Photograph One

Photograph One
Exercise: Outpost Creole 1
Platform: USNS Bartlett
Location: Entrance to Iceland-Greenland Gap
Equipment: NOSC HX231F sound source
Circumstances: After 2 days of seas in excess of 45 ft, the NOSC sound source team is inspecting damage to the sound source. We were unable to retrieve the source due to rapidly deteriorating conditions. Notice that the tail is bent approximately 45 degrees to the left. Vessel was under EMCON conditions and soon after a SAR P-3 flew overhead with LCDR McCallister as observer to ascertain our well being.

Photograph Two

Photograph Two
Exercise: Unknown
Location: Stern of an AGOR
Equipment: Bottom-mounted VEDABS
Circumstances: VEDABS IPV tied-down on the back of an AGOR; individuals shown deploying a bottom-mounted system

Photograph Three

Photograph Three
Exercise: Unknown
Location: Stern of an AGOR
Equipment: Bottom-mounted VEDABS
Circumstances: Bottom-mounted VEDABS being deployed from the stern of an AGOR

Photograph Four

Photograph Four
Exercise: Arctic Research and Environmental Acoustic Program (AREA) 85
Platform: USS Trepang (SSN 674)
Location: Canadian Basin - high arctic
Equipment: N/A
Circumstances: Trepang was one of the few submarines to actually surface near an EVA ice camp. She performed valuable services in conjunction with the measurements made at ice camps Opal, Crystal and Ruby.

Photograph Five

Photograph Five

Exercise: Outpost Heritage '86
Location: Chukchi Sea; Ice Camp Red
Equipment: Ice Camp Red
Circumstances: A photograph of Ice Camp Red.

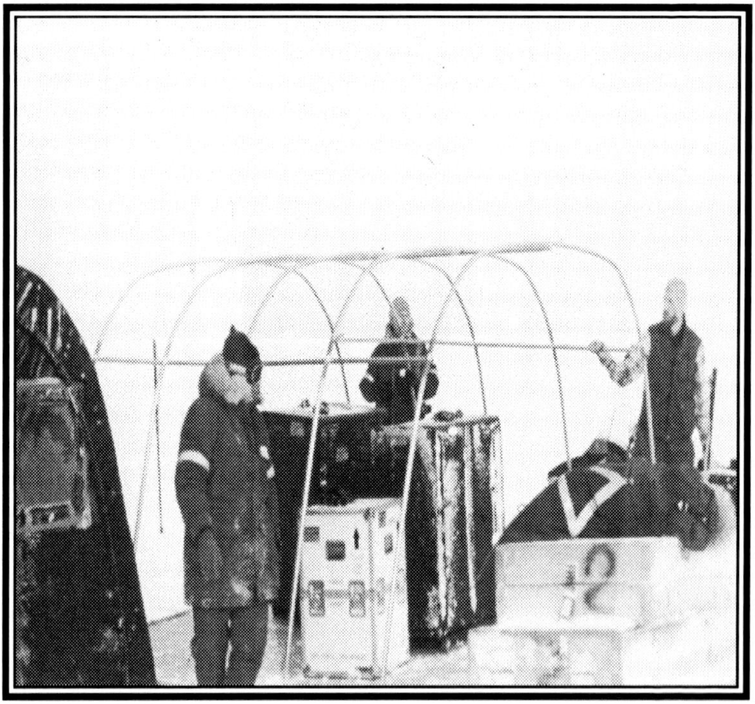

Photograph Six

Photograph Six
Exercise: ICEX '85
Location: Arctic Ocean
Equipment: Tent construction
Circumstances: Personnel constructing a tent where acoustic data will be recorded from the vertical, acoustic array which has been deployed through the ice in the tent next to the one being constructed.

Photograph Seven

Photograph Seven
Location: Chukchi Sea NW of Barrow Alaska (Ice Camp Rose)
Equipment: Photograph of a typical ice camp.
Circumstances: Separate tents are erected for acoustic measurements, environmental measurements, bunking and eating. Snowmobiles are standard equipment for transiting to and from remote array deployment locations.

Photograph Eight

Photograph Eight
Exercise: ICEX '85
Location: Arctic Ocean
Equipment: Ice agar
Circumstances: Personnel using an ice agar to make a hole in the ice for deploying an acoustic array.

Photograph Nine

Photograph Nine
Exercise: Outpost Heritage '86
Location: Chukchi Sea, Ice Camp Red
Equipment: Various/miscellaneous
Circumstances: Twin Otter shown was used to transport personnel and equipment to the ice camp.

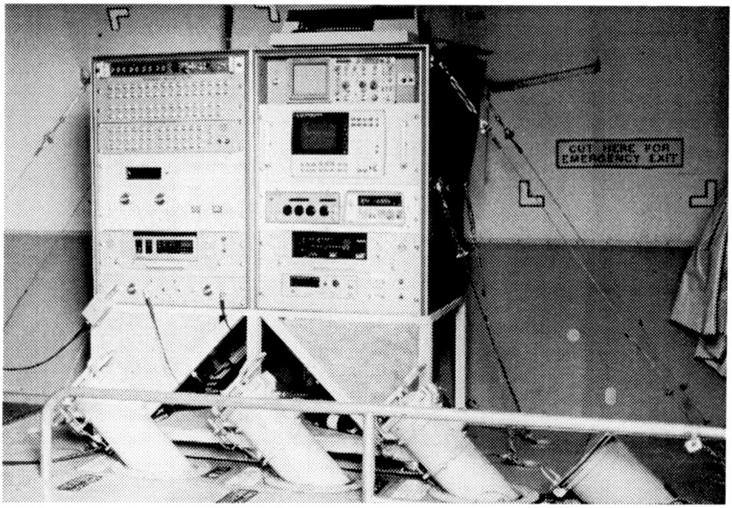

Photograph 10

Photograph Ten
Exercises: Encore, Marcot, ICEX 85 – 92, Broadband 87, CST-7, Glover
Platform: Various VP Squadron P-3 Aircraft and shipboard
Locations: JDF, Central Arctic, MIZ, Atlantic, Pacific
Equipment: POCO AMIGO (Portable Compact Acoustic Measurements Instrumentation for General Operations
Circumstances: AEAS Sponsored ARL:UT to develop and deploy airborne data collection system for quick acquisition of absolute sound pressure level data

APPENDIX D
Glossary

A/D conversion
Conversion of analog format to digital format.

ACODAC
Acoustic Data Capsule. A system to record acoustic signals.

acoustics
The physics of sound waves.

active sonars
Systems that can transmit acoustic signals. Generally they transmit acoustic energy and then receive the acoustic energy reflected from a target and the surroundings.

ADABS
An advanced ACODAC.

Admiral (ADM) Karl Dönitz
Leader of the German submarine force during WWII.

AEAS
ASW Environmental Acoustic Support. The successor to LRAPP.

AESD
ASW Environmental Support Detachment. A part of the Office of Naval Research, specifically formed to deal with computer models and their development.

ambient noise
All-encompassing sound at a given place, usually a composite of sounds from many sources near and far.

AMIGO
Acoustic Measurements Instrumentation for General Operations) airborne data collection system.

amplitude
Magnitude of the signal. Also called the signal strength.

AN
Ambient Noise. A term in the sonar equation.

anomalous detection
Detection which is not related to an identified source. The signal characteristics are stored for possible future identification.

Antisubmarine Warfare
The process of attempting to detect, and eventually have the ability to destroy enemy submarines.

armaments
Weapons.

arrival times
The specific times of arrival of a signal produced by an explosive charge, but due to the energy following different paths, the

signals arrive at slightly different times at a receiver.

Asdic
British term for sonar.

ASEPS
Acoustic Signal Excess Prediction System. A computer model.

ASN(R&D)
Assistant Secretary of the Navy for Research and Development.

ASRAP
Acoustic Sensor Range Prediction. A computer model.

ASTRAL
ASW Transmission Loss. A computer model.

ASW
Antisubmarine Warfare.

ASW prediction area
These were areas throughout the world that had homogeneous acoustic properties. These included water mass properties (a single sound velocity profile), constant water depth, and a single bottom loss value.

AT&T
American Telephone and Telegraph. The telephone company for many years.

Atomic Bomb
A bomb originally developed by the United States during WWII. It is based upon the principles of nuclear fission (splitting the atom). Used to end the Pacific War against Japan in 1945.

azimuthal variation
Angular variation of a property in the horizontal plane.

ballistic trajectory
Trajectory (path) followed by a body that is only affected by gravitational attraction.

band
A range of frequencies.

bandwidth
The size of the band.

basins
Geographical ocean areas, i.e., The Levantine Basin in the general area of the Levantine Sea in the Mediterranean off the coast of Italy.

bastions
Regions that are entirely within a nation's home waters. Also, the area where the nation has primary influence. For example, the

area where the Soviet Union principally controlled the ocean.

bathymetry

The characterization of the ocean bottom, generally portrayed as a chart.

bathythermographs

Depictions of the temperature variation of the ocean as a function of depth.

battery power

Power which comes from batteries.

Battle of the Atlantic

The struggle for the control of the North Atlantic Ocean during WWII.

beam

An angular measure of the sensitivity of a receiver or transmitter in a particular direction. Omni-directional beam means that there is no variation of strength or sensitivity as a function of angle.

beam width

The width of a beam either of a transmitter or a receiver. The width is generally defined as where the power has fallen to half of the maximum.

beamformer

Originally, a mechanical device that could form beams. In later years, the mathematical instructions in a computer on how to process the signals from different sensors.

BEARING STAKE

An exercise in the Indian Ocean. It is considered part of the LRAPP exercises because traditional LRAPP personnel were used to support the exercise. It was sponsored and support by PME124.

Bell Laboratories

The scientific part of AT&T, located in Whippany New Jerssey.

BIG DIPPER

A major LRAPP exercise, held in the Mediterranean.

biological reflectors

Living organisms, including fish and shrimp, that reflect acoustic signals.

biologics

Living organisms that generate noise. A major source of biological noise. Referred to as “snapping shrimp.”

BLUG

Bottom Loss Upgrade. A computer model that gave values of

bottom loss of acoustic signals that struck the bottom. The calculations were based upon the physical characteristics of the bottom itself.

bottom loss
The loss of acoustic energy suffered by an acoustic signal that strikes the bottom.

broadband
A band including many frequencies, as opposed to a single frequency.

CDC
Control Data Corporation. Makers of extremely powerful computers, used by FNWC.

characteristic dimension
The dimension (length) of a body. For example, the characteristic dimension of the Atlantic Ocean is approximately 3000 nautical miles.

Chief of Naval Operations (CNO)
The military head of the United States Navy.

CHURCH ANCHOR
A LRAPP exercise in the Pacific Ocean.

CHURCH STROKE
A LRAPP exercise in the Pacific Ocean.

classification
The type of submarine, based upon its acoustic signature.

classify
The process of deciding whether a received acoustic signal is coming from a submarine or not. If yes, which type of submarine.

climatological database
The historical data base of an environmental parameter. For example, the climatological sound velocity data base is the collection of all sound velocity profiles.

CNM
Chief of Naval Material. This office has been abolished.

CNR
Chief of Naval Research. The Commanding Officer of the Office of Naval Research.

Code 408
A particular code (office) within the Office of Naval Research.

Cold War (1946–1989)
The struggle between the Communist World and the Free World. Characterized by a lack of actual combat.

community, academic

The people who work for or are associated with universities.

community, oceanographic

The people who focus their professional efforts on problems and issues associated with the oceans.

community, scientific

The people who work as professional scientists.

Competition in Contracting Act

A federal regulation about the details of how federal contracts must be awarded. Organizations that work for the federal government must compete with each other in order to win the contract.

computer models

A set of instructions in a computer that take input data and manipulate it in order to produce desired information.

convergence zones

Regions where acoustic energy becomes concentrated. These regions occur due to the variation of the sound with depth, causing the acoustic energy to travel through the ocean in curved paths. They are characterized by a sharp drop in the transmission loss a signal suffers with increasing range from the acoustic source. They generally occur only in deep water.

convoys

An agglomeration of ships that move together for safety.

cruisers

A type of surface ship. They are larger than destroyers, but smaller than battle ships.

current shear

A phenomenon in a current where the ocean moves at different speeds over a relatively short distance, giving rise to shear.

DARPA

The Defense Advanced Research Projects Agency. Originally ARPA, or the Advance Research Projects Agency. Part of the Department of Defense.

data analysis plan

A very detailed plan on how to analyze the data obtained during a measurement exercise.

dB

A decibel. It is a tenth of a Bel. It is a mathematical function.

deep

In underwater acoustics, it generally means below the thermal layer which is near the surface of the ocean. It also means the operating depth of a submarine, as opposed to shallow.

depth charges ("ash cans")

Weapons dropped, generally from surface ships, that try to explode in the near vicinity of a submarine, causing fatal damage.

designations, money

The name for different uses of money within the Department of Defense. Some are for Operations, some for Research and Development, etc.

destroyer screens

Distribution of destroyers around an aircraft carrier for the purposes of protection.

detect

Sense a signal that could be come from a submarine.

detection

The process of detecting a signal.

detection range

The range or distance of the source of the signal from the receiver making the detection.

detection problem

A description of the issues associated with making a detection of a submarine.

DI

Directivity Index. A term in the sonar equation.

diesel engines

Engines that use compression to ignite the fuel and air mixture. Principal means of power for early submarines during WWI and WWII.

direct path

The path between a source and receiver which does not strike either the ocean bottom or ocean surface.

directional noise

Noise that comes in different magnitudes from different directions, generally in the horizontal.

directional

Not uniform in all directions.

Director of Antisubmarine Warfare Programs

OP095.

DSP

Digital Signal Processing.

EASTLANT
A LRAPP exercise in the Atlantic.

eddies
Large spinning masses of water.

electromagnetic waves
Energy propagating as waves in a particular part of the spectrum of frequencies. An example is radar waves.

electromechanical properties
Properties of materials. Mechanical deformation of the material gives rise to electric voltages. The application of voltage will cause the material to deform. These materials are used to make hydrophones.

environmental data
Data associated with the environment. This may be temperature of the ocean as a function of depth, the bottom contours, the state of the ocean surface, etc.

exercises, field
A coordinated set of measurements made in the field as opposed to the laboratory.

exercises, oceanographic
Exercises conduction in and on the oceans.

exercises, scientific
Exercises with their principal objectives being the measurement of scientific data.

exercises, fleet
Exercises by the operating fleet, using their equipment, as part of their training.

FACT
Fast Acoustic Coherent Transmission. A computer model used to prediction transmission loss.

filter
A device that will allow only a specific part of the signal to pass through. Characterized by a bandwidth. Some filters are wide (octave) and others very narrow (less than 1 Hz.).

filters, bank of
A set of filters, generally arrange in parallel (all signals being affected simultaneously) used to allow certain parts of the signal to pass through.

fire-control solution
A mathematical set of instructions that are organized and automatically transmitted to the weapon which tell it (generally a

torpedo in ASW) where to go in order to hit the target. Based upon inputs from the sensor suite of the operating platform.

firing solution

A fire control solution.

Flaming Datum

A phrase from WWII, indicating a ship that had been torpedoed and was generally on fire and sinking.

FLIP

Floating Instrument Platform. A very unusual ship that could either be in a horizontal position for towing, or in a vertical position for making measurements at sea.

FNWC

Fleet Numerical Weather Central. Also, Fleet Numerical Oceanographic Center.

force-in-being

A concept of an operating fleet with diverse units and capabilities.

Free World

Those nations that were in active opposition to the Soviet Union.

French Atlantic ports

Those French ports on the Atlantic coast used by the Germans after the defeat of France during WWII. The major port for submarine operations was St. Nazaire.

frequency

A measure of a function periodic in time, called a sinusoid. The number of repetitions per second is the unit. The number of cycles or repetitions per second is called a hertz (Hz). For example, 50 Hz means the signal is repeated 50 times per second, or 50 cycles per second.

Gaul, Dr. Roy

The Program Manager of the LRAPP program during the period 1971-1979.

great-circle route

The path which is the shortest distance between two points on the surface of a sphere.

Greater East Asian Co-Prosperity Sphere

The political organization desired by the Japanese Empire during WWII.

Greenwich Mean Time (GMT)

The time at Greenwich, England. The site of zero longitude. All navigation is based upon the time at Greenwich. Also called Zulu Time in the military.

hand-off problem

The problem of transferring the responsibility of observing a target from one platform to another.

heat

Energy transferred by a thermal process, specifically to do a different in temperatures. Heat may be detected by infra-red sensors.

Hersey, J. Brackett , Dr.

The first manager of LRAPP. Accomplished scientist, and noted leader of the oceanographic community.

HITS

Historical Temporal Shipping. A computer model of ships distributed over the ocean surfaces of the world Historical Temporal Shipping. There distribution is seasonally dependent.

hot spots

Areas in the ocean where the temperature is slightly higher than the surroundings.

Hot Wash-Up

A meeting that takes place immediately after an exercise. Information is exchanged, stories told, data transferred to the responsible parties.

hunter-killer

A group of USN ships operating off the coasts of the United States during the early days of the Cold War. They were looking for Soviet submarines.

hydrophones

Underwater sensors that could send and receive acoustic signals.

Hz

Hertz. A measure of the frequency.

initial velocity

Beginning speed.

initial angle

Beginning angle.

intelligence

Information about issues of interest. Intelligence about Soviet submarines Included the technical details of their construction, operations, etc.

internal computer

A computer internal to a particular piece of equipment. In the beginning of the Cold War, there were virtually no computers, and those that existed were huge, and required special handling. Today, there are computers internal to all types of equipment in both the military and civilian world.

ITASS

Interim Towed Array Surveillance System. One of the first towed arrays developed by the United States Navy.

Japanese merchant traffic

Japanese merchant ships operating between the areas controlled by the Japanese during WWII.

Jezebel

The original name for SOSUS.

kill

The process of destroying an enemy submarine.

knots

A unit of speed. One knot is One nautical mile per hour. One nautical mile is 6000 feet.

LAMBDA

Large Acoustic Marine Basic Data Array. A very long acoustic array, developed by DARPA. The original concept of Dr. Henry Aurand, who was the immediate predecessor of Dr. Roy Gaul in the operation of LRAPP. The array was over 3 nautical miles in length.

land warfare

Warfare operations that take place on land.

lines

Depiction of single frequencies. A line corresponded to a single frequency or tonal.

Lloyd mirror surface image interference

An interference pattern that arises when a wave train interacts with the reflections from the surface. Different frequencies cause different interference patterns.

localization

The process used find the position more accurately in a given area.

localize

The process of making a localization.

LRAPP

Long Range Acoustic Propagation Project. Administratively part of the Office of Naval Research, but having as a sponsor, OP095.

Manhattan Project
The project which developed the Atomic Bomb in the United States.

marginal ice zone
The region separating the major ice pack from open water in the Arctic.

Martell, Charles B., VADM
First Director of Antisubmarine Warfare within OPNAV. He was the original OP095.

master timeline
The chronological series of events that were to take place throughout an exercise. This included all ships, aircraft, and any participants and operations associated with the exercise.

Maury Center
The home of the oceanographic study center. LRAPP was one of the residents. The center was located on the grounds of the Naval Research Laboratory. It was known as Building 58. Named for Lt. Matthew Fontaine Maury, who was the first US Naval officer to scientifically investigate the oceans.

meteorologists
People who study meteorology.

MFA
Mid Frequency Array. A towed array specifically designed to operate at 300 Hz.

mine warfare
Warfare associated with the development, placement, and clearing of mines, particularly in the ocean.

missiles, anti-ballistic
A missile that is used to destroy ballistic missiles.

missiles, ballistic
A missile that once started, and its engines halted, follows a ballistic trajectory.

missile, cruise
A missile that could adjust its course and speed in flight.

missile, defense
A missile used for defensive purposes, generally against incoming missiles.

missile, incoming
A missile fired by the enemy and coming in as a threat against defense forces.

model bake-offs
: A comparison between computer models in a semi-public forum. Used to display and examine the capabilities and possible defects of possible future computer models.

models
: Computer models.

money, category 6
: Money used to support Research and Development efforts.

multi-engine aircraft
: Aircraft used by the Naval aviation community. Originally there were P2s, and then a series of P3 aircraft. The P3C was the Lockheed Orion aircraft. They were four-engine turboprops, used for anti-submarine and maritime surveillance.

multi-mission ships
: Ships which had capabilities to perform many different missions.

NADC
: Naval Air Development Center in Warminster PA.

National Command Authority
: The US President and the Secretary of Defense or their duly deputized alternates or successors. The NCA holds nuclear weapons release authority for the US armed forces.

NAVAIR
: The Naval Air System Command.

NAVELEX
: The Naval Electronic System Command.

NAVOCEANO
: The Naval Oceanographic Office.

NAVSEA
: The Naval Sea System Command.

Navy Standard models
: Computer models that have been both verified and validated for specific uses by senior officials. Dr. Hersey was the modeling coordinator of the U.S. Navy for oceanographic models. He was responsible for development and certification of Navy Oceanographic Standard models.

non-acoustic
: Those areas of physics that are not acoustics.

NORDA
: The Naval Ocean Research and Development Activity in Bay St. Louis, MS.

NOSC
The Naval Ocean System Center in San Diego, California.

nuclear warheads
Nuclear explosives that were attached to missiles.

nulls
Directions of substantial loss of sensitivity in a beam pattern.

NUSC
The Naval Undersea System Center in New London Connecticut.

ocean surface
The surface of the ocean.

ocean anomalies
Unusual variations of a property in the ocean.

ocean bottom
The bottom of the ocean.

Oceanographer of the Navy
A Rear Admiral responsible for all meteorology and oceanography in the United States Navy.

oceanographers
Professionals who study all aspects of the world's oceans.

ODSI
Ocean Data Systems Incorporated. An industrial organization.

omnidirectional
No preferred direction.

ONR
Office of Naval Research. Responsible for basic research and exploratory development in the United States Navy. The Commanding Officer is the Chief of Naval Research.

OP-095
The office of the Director, Antisubmarine Warfare. It is an office that is part of the CNO office, which is OPNAV.

operations analysis
Methods for investigation and analyzing operations

OPNAV
The Operations, Navy. It is the administrative organization that works directly for the Chief of Naval Operations.

optical waves
The waves that are in the optical frequencies. This is the portion of the visual spectrum.

optimum routing
Routes to be followed by ships and aircraft that will result in minimum fuel, or distance, or potential harm to equipment during

the course of travel from one point to another.

P2

A multi-engine aircraft used by the Navy for different missions.

P3

A multi-engine aircraft used by the Navy for different missions.

PARKA

A LRAPP exercise in the Pacific Ocean.

passive listening

Listening with a receiver for sound signals.

passive sonar equation

The equation which demonstrates the balance of different acoustic effects.

PE

The Parabolic Equation. A computer model.

physical oceanographers

Oceanographers interested in the physics of the ocean.

pig-boats

A name for German submarines during WWI.

pinged

Transmission of burst of acoustic energy from an active sonar.

Planning, Programming, and Budgeting System

A detailed management system begun within the Department of Defense under Secretary McNamara. The acronym is PPBS.

platforms

A term used in the Navy to describe major types of equipment. There was a surface platform (surface ship), a submarine platform (submarine), etc.

PME124

An office within NAVELEX. This office was responsible for all Undersea Surveillance. It was responsible for SOSUS and the development of SURTASS.

probability of detection

The chance of detecting a submarine. This was calculable and depended upon values obtained from the sonar equation.

Program Manager

The individual responsible for all phases and components of a defined program.

radar

An active system, invented prior to WWII. Radar sent out a burst of electromagnetic energy, and then detected the returning reflection from a target.

radar, airborne
A radar system carried and operated aboard aircraft.

radar detection system
A system used to detect targets using the physics of radar.

radar, surface ship mounted
A radar mounted on board a surface ship.

range-dependent variation
Variation of a variable with distance. Generally the distance was considered to be in the horizontal direction.

ranges, operating
Distances where the operation of a system is viable.

RDT&E
Research, Development, Test and Evaluation.

reacquire
Find a target, again. Frequently used in the context of the hand-off problem.

receiver
The part of the system that detects and organizes the incoming signal.

receiving sensors
The sensors that detect the incoming signal.

reflected echo
The signal that is reflected from a target and then detected by the receiver.

reflection
The process of be re-transmitted after interaction with a target.

reflectivity
Characterization of the strength of the reflectivity of a target due to its material and geometric shape.

refract
To bend as the acoustic wave moves through the medium. This occurs because of the variation of the properties of the medium orthogonal to the direction of propagation of the signal.

repeaters
Devices that will amplify a signal as it travels along an underwater cable.

RFP
Request for proposal.

RP70
One of the first long range acoustic propagation prediction computer models.

S/N

Signal to Noise ratio. Written in this form, it generally implies a logarithm.

salinity

Salt content of the water.

SEA SPIDER

A LRAPP experiment in the Pacific Ocean.

seamounts

Mountains rising from the ocean floor.

sensors

Devices that can sense physical phenomena.

shallow

In underwater acoustics, it generally means above the thermal layer which is near the surface of the ocean. It also means the operating depth of a submarine, as opposed to deep.

SHARPS

Ship Helicopter Acoustic Range Prediction System. A computer model that made predictions for the fleet, operated at FNWC.

Shear, Hal, VADM

Second OP095. Sponsor of EASTLANT.

Shipping surveillance

Examination of the ships at sea. Originally from aircraft. Now it can be accomplished by satellite. Detailed information about a particular ship is called “rigging” and means flying close to the ship and taking photographs and other pertinent measurements.

ships, merchant

Commercial surface ships that carry cargo of all types across the oceans.

sidelobe

A beam pattern has a main lobe, which is characterized by the maximum response of the system in a particular direction. In a direction other than the main beam there are other local maxima. These are called the side lobes. The side lobe level is a measure of the reduced sensitivity compared to the main beam.

signal strength

The amplitude of the signal.

signal-to-noise ratio

The ratio of the strength of the signal of interest to the strength of the noise. In the logarithmic form it is generally written as S/N.

signatures

Characterization of the frequency content of the signal from a source. The signature may have no lines, or several lines. The characterization is the frequency of the lines and their relative amplitudes.

SL

Source Level. A term in the sonar equation.

snorkel

An air supply tube that would allow air to be brought into a WWII German diesel submarine while the submarine remained just under the surface of the water.

SOA

Speed of Advance.

sonarman

At sea, the operator of a sonar on board a ship or submarine.

sonobuoy

A device that could be dropped into the water and then deploy a hydrophone that would sink to a specific depth. It would transmit its received signals by radio to generally an aircraft or a surface ship.

SOSUS

Sound Surveillance System. It was a series of hydrophone arrays, lying on the bottom of the ocean, and cabled back to a shore station using underwater cables and repeaters. It was the fundamental system that detected submarines operating in the deep oceans throughout the Cold War.

sound velocity profile

The variation of the speed of sound with depth, measured from the ocean surface to some depth, and sometimes, to the bottom.

sound propagation

The act of movement of the acoustic signal from one place to another within the surrounding environment, generally the ocean.

sound velocimeters

Equipment specifically made that will directly measure the speed of sound in the ocean as a function of depth.

sound channel

The channel that is formed by the variation of the speed of sound with depth from the surface value to the minimum value and then increase to the same value it had at the surface.

source levels

The strength of the source of energy.

Soviet Command Center

The place in Moscow where the military leaders of the Soviet armed forces meet to control their troops.

SOW
Statement of Work. Part of a contract specification.

spectra
Plural of spectrum.

spectrum
The amplitude of the energy emissions of a source, as a function of frequency.

spectrum, radiated noise
The spectrum of the noise radiated by an acoustic source.

SPPS
Early form of ASEPS. A computer model.

SQS-23
A surface ship sonar.

SQS-26
A surface ship sonar with the ability to steer its beams, and also to bounce the acoustic energy off the bottom. A successor to the SQS-23.

SQUARE DEAL
A LRAPP exercise in the Atlantic Ocean.

St. Nazaire
A French port on the Atlantic coast, used by German submarines.

stationary
Time independent. No properties change with increasing time.

stealth platform
A platform that cannot be detected, except with great difficulty.

submarines
Ships designed to operate below the surface of the sea.

submarine, attack
A submarine designed to attack surface vessels and other submarines.

submarine, diesel
A submarine powered by diesel engines on the surface and batteries while submerged.

submarine, enemy
Not one of ours.

submarine, missile-carrying
A submarine designed to launch missiles. Originally the missiles were ballistic. Now they are cruise missiles.

submarine, nuclear
A submarine powered by a nuclear reactor.

submarine tender
A surface ship specifically built and stored with equipment to support all requirements of a submarine.

submarines operations center
The location where the control of all the submarines in a particular area is exercised. The central point for all communications.

submersible surface ship
The characterization of early submarines.

supply life line
The paths taken between the United States and England by merchant ships carrying goods and sustenance during WWII.

surface warships
Warships that operate on the surface of the sea. An example is a destroyer.

SURTASS
A towed array system, including the towed array and the surface ship to pull it, developed by the United States Navy.

surveillance system
A system that can observe the operations of the enemy. In this context, it is a system that can observe the movements of enemy submarines. It consisted of SOSUS, towed arrays, and other equipment.

SUS
Source Underwater Signal. An explosive device that could be dropped from ships and aircraft. When it exploded at a pre-set depth the acoustic signal was received by different receivers placed throughout the exercise area.

SYSCOMS
System commands. This included NAVAIR, NAVELEX, and NAVSEA.

tactics
Operations designed to produce a particular result.

targets
Generally enemy submarines, but any enemy naval vessel that could be detected, and possibly destroyed.

Task Force 67
An operational component of the Sixth Fleet, which operated in the Mediterranean Sea.

TASSRAP

A LRAPP exercise that took place in the Mediterranean. Also, the name of a computer model.

TEST BED

A LRAPP exercise that took place in the Atlantic.

thermocline

The characterization how the water temperature varies with depth.

threat

An enemy platform that can threaten United States interests. In this context, it is generally a Soviet submarine.

time series

The values of a variable generated successively in time. The series may be continuous or discrete.

TL

Transmission Loss. A term in the sonar equation.

tones

An expression of a specific frequency.

torpedo

A device used to sink ships. Launched from a platform (submarine, surface ship, or aircraft) it proceeds independently along a specific course until it hits or misses the target. Has a large non-nuclear warhead with conventional explosive. Could be considered to be an underwater missile.

total noise

The aggregate of all the noise at a particular point.

Towed arrays

A primary detection device of Soviet submarines starting in the early 1960s. This was generally a long hose populated with hydrophones, separated at carefully controlled distances. The array, commonly called a *tail*, was towed at low speeds, generally below 5 knots. These arrays were used by both surface ships and submarines.

trail

The tactic used by US submarines in following Soviet submarines at close distances. The US submarine was usually above and to the side of the Soviet submarine. This maneuver was very dangerous, and had the inherent risk of collision. The trail was first achieved through the process of detection, classification, and localization. It also included several hand-offs. The trail generally lasted throughout the voyage of the Soviet submarine.

turbo-generators
Turbines used to generate electricity.

Type XXI
The most modern type of WWII German diesel submarine. The war ended before they could be used in quantity.

U-boats
German submarines during WWII.

U.S. Naval Postgraduate School
A graduate school maintained and run by the United States Navy for special training. Located in Monterey California.

undersea warfare
Warfare taking place in the ocean depths.

underwater acoustics
Application of acoustics to underwater systems.

underwater acousticians
Professionals who study acoustics for applications in the ocean.

underwater sound propagation
Transmission of acoustic signals in the ocean.

unrestricted submarine warfare
A policy used in both WWI and WWII. All ships at sea were considered fair game for the German submarines. No warning was given before an attack.

USS Holland
The first United States submarine, SS-1, developed by John Holland.

USS Nautilus
The first nuclear submarine of the United States Navy.

validation
A process of subjecting a computer model to rigorous review. Comparison of predictions to measurements. Once it is validated a computer model may be used for a variety of functions within its boundaries of validity.

VEDABS
Versatile Experimental Data Acquisition Buoy System. An extension of the ACODAC, it was the primary acoustic measurement system used by the AEAS Program from 1982 to about 1993.

velocity profile
Sound velocity profile.

verification

A process of subjecting a computer model to rigorous review. Careful scrutiny is made to make sure that the computer code is actually doing what the designers of the model wished.

VP

Multi-engine aircraft.

war games

Realistic exercises where a scenario is followed which allows the participants to actually use all their military skill and equipment to accomplish the objectives of the exercise. This may involve thousands of people and last for several weeks.

water mass boundaries

Boundaries of a water mass which had consistent properties throughout.

wavelength

In general, the mean distance between maximums (or minimums) of a roughly periodic pattern. The frequency and wavelength are inversely proportional. The product of the frequency and the wavelength equals the speed of propagation. In water, a frequency of 50 Hz has a wavelength of approximately 30 meters.

weapons

Armament. In this context, it is generally torpedoes and missiles.

Western Electric Company

The manufacturing arm of AT&T.

white noise

Noise for which the energy density is independent of frequency over a specified bandwidth.

Wilhelmshaven

The German Baltic port where the German submarines originally operated from before the fall of France in WWII.

Winston Churchill

Prime Minister of Great Britain during WWII.

Wolf Packs

A collection of German submarines, generally operating on the surface, that simultaneously attacked allied convoys in the beginning of WWII.

Woods Hole

The Woods Hole Oceanographic Institution, located in Woods Hole Massachusetts.

World War I

The armed conflict between the Allies and the Axis during the period 1914-1918.

World War II

The armed conflict between the Allies and the Axis during the period 1939-1945.

Zulu Time

Military term for Greenwich Mean Time.

APPENDIX E
LRAPP Community

LRAPP Community

Last Name*	First Name	Initials
Adams	Budd	B.B.
Adams	Dick	R.M.
Adler	Bob	R.
Akers	Morris	M.R.
Anderson	*Aubrey*	*A.L.*
Anderson	George	G.B.
Anderson	Vern	V.F.
Andrulis	Marni	M.W.
Arndt	Larry	
Aurand	*Hank*	*H.S.*
Baggett	*Lee*	
Baker	Pat	
Ball	Tracy	
Barbera	Jim	J.G.
Barnes	Alan	A.
Bedford	Nancy	
Bell	Bobbie	C.R.
Bernard	*Phil*	*P.G.*
Bissett	Ed	
Blue	Joe	
Blumenthal	Barry	B. P.
Bossard	Dave	
Boutin	Paul	P.R.
Bowen	Julie	J. I.
Bowman	Chuck	C.
Bradley	Dave	D.L.
Bradley	Marshall	M.
Brancart	Claude	C.P.
Brookes	Al	A.G.
Brown	Roy	R.T.
Browning	David	

**Names in Italics are deceased*

Brunson	Burlie	B.A.
Bucca	Paul	P.J.
Bucker	Homer	
Buck	Beau	
Buck	Manor	
Caldwell	*Turner*	
Cann	Gerald	
Carey	Bill	W.M.
Cavanagh	Ray	R.C.
Chaika	Ed	
Chwastyk	Abe	A.M.
Clark	Tony	
Clausner	Ed	
Coblentz	Esther	
Colborn	Joe	
Cole	Bernie	B.F.
Collins	Joe	
Corwyn	John	
Cybulski	*John*	
Dabney	Chuck	C. H.
Darrell	Charlie	C.G.
Daubin	*Scott*	*S.C.*
Davis	Chuck	
Davis	*Ed*	*E.E.*
Davis	Jim	J.A.
Davis	Vic	V. H.
Deleeuw	David	
Diachok	Orest	O.I.
Dial	Ken	
Didier	Betty	
Diloreto	Gene	
Doblar	Randy	
Dontaler	Lee	

**Names in Italics are deceased*

Doolittle	Dick	R.D.
Dorman	Craig	C.E.
Dunbar	Bob	R.D.
Dunn	Jeanie	
Durham	Don	D.L.
Early	*Paul*	*P.J.*
Edelblute	Dave	
Ehlers	John	
Elam	Sam	S.R.
Eller	Tony	
Ellis	Glen	G.E.
Eppert, Jr.	Herb	H.C.
Evans	Kirk	K.E.
Fadness	Bud	
Feden	Bob	
Feldmann	Peggy	
Fenner	Don	
Feuillet	J. P.	J.P.
Fisher	Fred	F.H.
Foxwell	Dick	
Franchi	Ed	
Frosch	Bob	R.A.
Gaffney	Paul	P.G.
Gagney	Lou	
Garabed	Ed	
Gardner	Bob	R.R.
Garon	Howard	
Gaul	Betty	B.J.
Gaul	Roy	R.D.
Gereben	Steve	I.B.
Gholson	Norm	
Golden	Joe	
Goodman	Ralph	R.R.

**Names in Italics are deceased*

Gotthardt	Jerry	
Gottwald	Jim	J.T.
Gough	Ed	
Graham	Elwyn	E.D.
Haas	Paul	
Hampton	Loyd	L.D.
Hanna	John	J.S.
Harkness	John	J.R.
Harned	Winky	
Harris	Mike	
Hashimoto	Eigo	
Hasse	Ray	
Hawker	Ken	K.E.
Hayes	*Earl*	*E.E.*
Hecht Dick	R.J.	
Henderson	Russ	
Hersey	*Brackett*	*J.B.*
Herstein	Peter	
Hess	John	
Hoffman	Jerry	
Hogg	James	
Jackson	*Dempster*	
Jackson	Roy	R.V.
Jacobs	*Gil*	*G.V.*
Jeanette	Bob	
Keen	Jean	D.J.
Kelley	Fred	F.J.
Kirby	Bill	
Kirklin	Dick	R.H.
Kirst	Al	A.
Klinkner	Dick	
Koenigs	Paul	P.E.
Kuhn	Ira	

**Names in Italics are deceased*

Kulek	Sid	S.
Lackie	Joanne	
Lackie	Skip	K.W.
Lambert	Peggy	P.A.
LaPlante	Bob	R.F.
Larsen	Ron	R.W.
LaSalle, Jr.	Don	
Latham	Dick	
Lewis	Glynn	M.G.
Lynch	Len	L.T.
Mackenzie	Ken	
Malahoff	Alex	A.M.
Marshall	Sam	S.W.
Martel	*Charles*	
Martin	*Bob*	*R. L.*
Matheny	Cal	C.W.
McAllister	Mike	
McCaffrey	Joe	
McCammon	Diana	
McCloskey	Terry	T.J.
McDermid	Jack	
McGirr	Bob	
McKee	Kinnaird	
Milburn	Darrell	
Mitchell	Steve	S.K.
Morenoff	Ed	
Morey	Carl	C.F.
Morningstar	Chuck	
Morris	Jerry	G.B.
Morrison	Bob	
Moseley	Bill	W.B.
Muellenhoff	Bill	W.P.

**Names in Italics are deceased*

Murphy	Daniell	
Muzzey	*Charlotte*	
Nardello	John	
Newton	George	
Northridge	Bruce	
Nowak	Dick	
Osborne	Kenneth	K.R.
O'Sullivan	George	
Paoletti	Chuck	
Paquin	James	J.E.
Pedersen	*Mel*	
Penrod	Clark	C.S.
Perrone	Tony	G.A.
Pickens	*George*	
Pittenger	Dick	R.F.
Raisbeck	Tobey	G.
Ramsdale	*Dan*	*D.J.*
Rather	*Roy*	*R.L.*
Reeves	Jesse	
Rojas	Dick	R.R.
Ronaldi	Tom	
Saalfeld	Fred	F.E.
Salinas	Manny	
Santinello	Sal	
Schenck	Harry	
Schmid	John	
Schreiber	Bill	W.G.
Sharp	Jim	J.M.
Shear	*Hal*	
Shooter	Jack	J.A.
Shultz	Glen	
Simmons	Vern	
Smith Ed	E.L.	

**Names in Italics are deceased*

Smith Bob	R.	
Snodgrass	Jim	J.M.
Snyder	Ed	J.E.
Solomon	Louis	L.P.
Spansel	Mira	
Spiesberger	John	
Spiess	Fred	
Spikes	Clay	
Spindel	Bob	R.C.
Spofford	Charles	C.W.
Stanbrough	*Jess*	*J.H.*
Stewart	Jim	
Stuart	Ann	A.H.
Stuart	Charles	C.E.
Sullivan	Don	
Swenson	Rick	R.C.
Sykes	Wendell	W.G.
Tatro	Peter	P.R.
Thorp	Bill	
Tunstall	Ed	E.B.
Urick	*Bob*	
Van Orden	Dick	M.D.
Van Schuyler	Pete	
Von Winkle	Bill	
Wagstaff	Ron	
Waller Ed	E.C.	
Watrous	Ben	
Weinberg	Chic	
Weinstein	Marvin	M.S.
Wheatley	Bobby	B.N.
White	Dewayne	
Williams	Budd	W.J.
Wilson	Jim	

**Names in Italics are deceased*

Winokur	Bob	R.S.
Wittenborn	*Gus*	*A.F.*
Wolff	Paul	
Worsely	Wayne	
Yee	Grant/Leila	
Young	Ernie	E.T.

**Names in Italics are deceased*

APPENDIX F
The Basics

General Comments

This book is intended for people who are interested in a special part of the history of the Cold War. Part of this world-wide complex struggle was between the navies of both sides. This competition reached its highest level of intensity in the generic area of undersea warfare, with specific application to the area of Antisubmarine Warfare (ASW). While the struggle was fought by men and women on both sides, to an unusual extent the outcome was controlled by science. The understanding of the ocean environment was critical to the outcome of the struggle. This section introduces and explains the concepts of how the ocean environment affects the battle.

I am, however, obliged to say that readers who do not wish to familiarize themselves with the technical concepts that ruled the ASW struggle may skip this section. However, for the reader who wishes to appreciate the efforts and the de-tails of the struggle, this section will provide information that will clarify the efforts on both sides. Substantial effort has been made to keep the use of technical jargon or equations to a minimum.

The general problem of ASW has many parts, but at the most fundamental level, the problem for the ASW Forces is to *find the target submarine*. This process is called detection. Further parts to the process involve classification (is it a submarine, and if so, what kind of submarine), localization (precise location of the submarine), and the ability to kill the submarine (fire control solutions for the weapons used).

The detection process is dependent upon the noise generated by even the most quiet of submarines, the ocean environment and its effects upon the acoustic energy radiated by the submarine, and the characteristics of the receiver listening to the submarine. In essence, it is necessary to determine the geometry of the problem. In simple terms, what is the distance (range) between the source (the submarine) and the receiver (the sonar system)? This calculation for arriving at this information is based upon the sonar equation.

The Sonar Equation

At its most fundamental level, the technical story of ASW depends on understanding the sonar equation. While this calculation might appear to be forbidding, it is actually a single, rather short equation with only a few terms. Keep in mind, however, that this single equation drove the efforts of all the participants on both sides of the undersea warfare battle.

To illustrate the equation, imagine that we want to detect (using our senses) a signal source. The argument applies to underwater acoustics, astronomy, speech, or any other activity where there is a signal source, a path that the signal must travel, and something to receive the signal. In the case of a person talking in a room, the source is the speaker's mouth; the source level is how loud they speak. For the sake of clarity and without going into technical details, imagine that you are trying to listen to your friend, George, who is talking to you in a room. The derivation of the sonar equation can be accomplished using this pedestrian example.

The ability to hear George speak is dependent on several things, all which will be discussed in their appropriate time. The first issue, however, is to recognize that the *signal* you hear, i.e., George's voice, is dependent upon how loud he is talking, and how far away you are from him. If he speaks at a constant level, i.e., doesn't speak louder or softer, but keeps talking at a constant level (*source level* is constant), then the distance you are from him will be a major factor in affecting whether or not you hear him.

Now if you are very close to George, you can hear everything he is saying. The distance at which you hear him is known as the *detection range*.

When you can hear him about 50% of the time, that range is known as the *maximum detection range* . Now, begin walking away from him while he keeps talking. You will notice that as the distance increases your ability to hear him will diminish. In technical terms, the signal strength of his voice, as measured at your ear, has diminished. This reduction in the signal strength is known as *transmission loss*. In different terminology, the reduction of the signal strength is due to the distance from the source and other factors that have not been discussed, but will be introduced later. Thus, the received signal strength or the perceived strength of George's voice has diminished as you get farther away from him.

The signal arriving at your ear, the *receiver*, is now discernible down to a certain level, which corresponds to a certain distance. In other words, the amount of speech that you can understand while George is talking at a constant level is a *function of the separation distance*. In addition, your ear has certain characteristics that will allow it to discern some voices better than others.

I have said nothing about the other events going on in the room. If you are alone with George in a large room and can go as far away from him as you wish, there will be a certain distance where you can barely hear George. Sometimes you hear a few words, and sometimes hear none. If you move closer to George, then you will hear more words; if you move farther away from George, then you will hear fewer words.

Now, let us make the matter a little more complicated. Let us add talking people to the room. The problem remains the same: you want to listen to George, but all of a sudden, people who are talking to each other appear in the room. You have no interest in what they are saying. They are generating *noise*, i.e., they are not generating signal of importance to you. Noise can be defined as unwanted or interfering signal. The interfering speakers are throughout the room, and the noise they make is known as *ambient noise*.

Only the speaker generates signal. You immediately recognize that to hear only George, you will have to move closer to him. In technical language, the ambient noise reduces the detection range. It can be thought of as a struggle between the source level (strength of the speaker's voice) and the ambient noise (the sound from other people in the room talking).

If you want to continue to hear about half of what George says, there are several options. If you remain at one place, then he can talk louder (increase the source level), or you can ask the people who are talking to be quiet (reduce the ambient noise). If neither are under your control, then you will have to move closer to George (reduce the transmission loss).

There is another way that you can better understand George in the presence of noise without reducing the distance between you. You can get better ears. It is well known that young people (those who have not listened to loud music) have better hearing than older people; that is the physiology of hearing. But do not lose the point. If you could somehow point your ears in George's direction or in some way block out the noise that is due to the all the other people talking in the room, then you could hear George better. If this were possible, then you could improve your *detection threshold* or increase the *detection range* by improving the listening equipment (your ears). The level of the signal is the important measure. In technical terms, the *signal excess* is a measure of the received signal strength above the noise or interference. As the signal excess gets larger, you can understand more words; as the signal excess gets smaller, you can understand fewer words.

This information is technically stated in various forms of the sonar equation.[1] In layman's language,

Signal Excess=Source Level-Transmission Loss-Ambient Noise+ Array Gain

The case of signal excess = 0 (that is, you can understand about 50% of what George is saying) occurs when

Transmission Loss=Source Level-Ambient Noise+Array Gain.

This calculation is known as the *Passive Sonar Eqation*, and it contains four terms: *transmission loss*, *source level*, *ambient noise*, and *array gain*, all of which must be discussed in more detail. Both the transmission loss and the ambient noise are substantially affected by the ocean environment, i.e., the speed of the sound in the ocean, on the bottom, and at the surface of the ocean.

For most of the efforts in ASW during the Cold War, the passive sonar equation was used because the ASW systems were passive; they did not use an active source. The concept of active sonar implies an active source.

- Active sonar is where an active source transmits acoustic energy, which propagates through the medium, strikes and reflects off the target, and then returns to a receiver that may not be co-located with the source.

The active sonar equation is slightly more complicated, but that is not important for our purposes.

Transmission Loss

The transmission loss is purely a function of distance in a very large room (no reflection of the acoustic energy from the walls, floor, or ceiling). As the distance increases between the source (the speaker, George) and the receiver (your ears), the transmission loss increases as the square of the distance between them. This is referred to as spherical spreading.

The acoustic signal propagates at the speed of sound. That is, the medium (in this example, air) has a particular property that is called the

---

[1] For technical reasons, the Sonar Equation is expressed in logarithms. Instead of multiplication, you can use simple addition. The use of logarithms will generate both positive and negative numbers. This is merely a method of arithmetic and does not affect the fundamental discussion.

speed of propagation, or since it is sound that is being propagated, the *speed of sound*. In technical terms, it is the speed that a pressure disturbance travels through the medium. In air, the speed of sound is dependent upon temperature. In the simple example that we are using—a speaker and a listener separated by a specific distance—the signal will take a certain finite amount of time to propagate from George (the speaker) to you (the listener). If the air in the room becomes uniformly hotter, then that time will decrease, and if the air in the room becomes colder, then that time will increase.

As the pressure signal (the sound) propagates through the air, it changes character for several reasons, which are due to some complex phenomena, as well as hitting and bouncing off walls, people, or other objects. The details of why the signal character changes are not important in this discussion; but it is important to recognize that the character of the signal does change.

The point is that the characteristic of the signal is affected during its passage through the medium. This observation has enormous significance for the entire problem of propagation of sound underwater.

Speed of Sound in the Ocean

In ASW, the acoustic signals are propagated through the ocean environment. The ocean environment has some unusual characteristics that will affect how sound is propagated over any substantial distance. As in our example about the speakers and listeners in an air-filled room, the characteristics of the environment are important.

The speed of propagation in the ocean is dependent upon temperature, salinity, and density. It is important to remember that the ocean has three dimensions: latitude, longitude, and depth. At each spatial point on the surface of the ocean surface (sea level), the ocean has some depth. Any point in the ocean has an associated temperature, salinity, and density. Simply expressed, the speed of sound varies at each point in the ocean. For our purposes, it is convenient to think of the variation of the speed of sound with depth. The speed of sound changes with increasing depth for a fixed lateral position.

For simplicity, consider the speed of sound variation in three general areas of depth. Between the sea surface and a depth of approximately 100 meters (approximately 300 feet), the speed of sound is heavily influenced by temperature. During the day while the sun is out, the upper region warms; during the evening hours while the sun is hidden, the upper region cools. Thus, the speed of sound always varies to some extent in that region.

This region is also affected by wind, rain, and other local weather conditions.

As you go deeper in the ocean, the effect of temperature decreases. Sunlight does not penetrate into the deeper regions and the water is colder. The pressure also begins to increase substantially as you go deeper. Both the struggle between the warming and cooling effects of the sun, and the effects of pressure and its subsequent effect on density, combine to produce a minimum in the sound velocity.

As you travel still deeper, the effect of temperature vanishes completely, and the speed of sound is entirely dependent upon the pressure. In fact, the speed of sound increases linearly with depth in the deep ocean.

Because of the minimum speed of sound at some depth, the profile resembles a cup turned on its side. This minimum speed of sound, with higher speed of sound both above and below this depth, causes the sound to propagate as if it was in a channel. The depth where the speed of sound is a minimum is called the *sound channel ais* .

Since the depths below the sound channel axis are essentially independent of any environment that is happening close to or at the sea surface, that part of the speed of sound profile is invariant with season, or even geographical position on the face of the globe.

Now, our discussion of the sound profile has to be slightly extended. Clearly, the upper surface of the sound velocity profile will be different in the northern latitudes (e.g., the Arctic) from the upper surface in the tropics near the equator. This difference means that the sound velocity profile varies with both depth and horizontal position. However, the depth of the ocean is at most only a few miles, and the distance from the pole to the equator is thousands of miles.

Let's consider a particular large ocean region, say, the central North Pacific between the International Dateline and the West Coast of the United States, and from 20° N to about 30° N. The distances that characterize the area on the sea surface are thousands of miles, but the depth of the ocean is only a few miles. Although the speed of sound varies throughout the region, it changes slowly with lateral position, but comparatively rapidly with depth. (In mathematical jargon, the speed of sound changes rapidly with depth, but only slowly with range.)

The fact that the speed of sound varies in the ocean has a major effect on the path that the acoustic signals follow. The concept of a path is easy to understand, but the mathematical definition is more involved and requires some explanation. The sound that emanates from an acoustic source consists of a wave front: a mathematical surface across which properties can be related. Specifically, a wave front can be defined to be the surface that has a constant property.

Regardless of the path that has been followed, each point on the surface has the same property value. In the case of travel time, each point on the wavefront surface has the same value, that is, each part of the wavefront has traveled the same amount of time. All the waves emanated at the same start time (i.e., time=0) from the point source. If you draw a short line from each point on the surface that is normal to the surface, and then trace out the position of these normal vectors in small time steps, then you will generate paths. The lines (loci) you made are called the *ray paths*. It is easy to see that the angle to the horizontal is changing at each point along the path; this effect becomes important when we discuss sonar receivers.

Ray paths are straight lines in a constant environment (such as the room we used as an example of how sound travels). However, in such an environment as the ocean where the speed of sound is not constant, the ray paths are not straight lines: they curve. This one fact leads to some amazing consequences. Because the speed of sound varies with depth—first decreasing with increasing depth, then reaching a minimum, and then increasing with increasing depth—the ray paths are "squished" together and, in some cases, are spread apart.

This fact implies that the transmission loss varies with range from the source and is dependent upon the depth of the receiver. Further, it means that the transmission loss does not always increase with range. In fact, if the ocean is deep enough and if the surface is warm enough, the transmission loss suddenly sharply decreases for a short range (a few miles or less) and then continues to increase in some places. These regions are called *convergence zones*. The "deep ocean" occurs over most of the oceans throughout the northern hemisphere, once away from the relatively shallow regions near the coasts. In the deep ocean, convergence zones occur roughly 20 to 30 miles apart. The Pacific Ocean is different from the Atlantic Ocean, which is different from the Mediterranean Sea. The difference between the three major oceans is due primarily to the different character of the sound speed profiles. This difference is due to the variations in salinity, temperature at the surface, and bottom depth. Again, the key point here is that well-defined regions called convergence zones exist, and in these zones the transmission loss is sharply reduced.

Of course, the actual character of the convergence zones is substantially influenced by the source and receiver location. Deep sources with shallow receivers will have convergence zones associated with them that are different from the convergence zones associated with shallow sources and deep receivers.

Attenuation of Signals in the Ocean

Conceptually, we have been thinking of acoustic signals that have a single frequency. However, the ocean absorbs acoustic energy as a function of frequency, so some signals that are of exactly the same amplitude (signal strength or loudness) but have different frequencies will experience different transmission losses. Frequencies around 19–20 Hz incur little attenuation in the deep ocean. Whales communicate at that same frequency, which is how they can find each other over the enormous distances they travel.

Certainly, there is a direct correlation between the continued existence of whales and their ability to communicate, and, of course, their communication systems. Attenuation increases slowly for signals up to approximately 1000 Hz (1 kHz), and begins to increase rapidly with increasing frequency. So, for signals in the 100-kHz region, their amplitude decreases rapidly so that within a few miles the signal cannot be detected. For signals in the 50-Hz region, the signal can propagate for thousands of miles without substantial signal attenuation. For very low frequencies, such as 5 Hz the attenuation is even less, but the ability to detect signals is rather low. The key point is that the attenuation of the acoustic energy in the ocean is a function of frequency.

Bottom Interaction

The bottom of the ocean is extremely varied. It is rocky in some places and muddy in others. The shape of the bottom is called the *bathymetry* by ocean scientists. The mapping of the bottom bathymetry has continued for over 160 years.

The first efforts to understand the ocean and its bottom were begun by an American naval officer, Lieutenant (LT) Matthew Fontaine Maury, USN. His work, begun in approximately 1835, eventually resulted in the establishment of the Depot of Charts and Instruments, which eventually became the U.S. Naval Oceanographic Office (NAVOCEANO). His mapping of the ocean bottom and the accompanying data base of many oceanographic parameters, including the measurement of sound speed profiles, survive and continue to be used to this day.

When the acoustic energy leaves the acoustic source, in our case, a submarine, it propagates through the water as described previously. However, if the ocean is not quite deep enough or if the sound speed profile is of a particular shape, the acoustic energy hits the bottom (called *bottom interaction*).

In the science of oceanography, it was assumed for many years that when bottom interaction occurred, the acoustic energy was absorbed and simply disappeared. Now we know that is not the always the case. Under certain circum-stances, which are not particularly rare, the acoustic energy enters the bottom, and then exits at some distance later. The sound propagates in the bottom itself. Due to the increase of the speed of sound in the bottom compared to the water immediately above, the sound curves while in the bottom, and returns to the water some distance away. This curving of the sound waves is caused by exactly the same physics that we discussed earlier.

The bottom frequently consists of deep layers of sediment. This sediment has a sound speed associated with it. Generally, the speed of sound is much higher than the speed of sound in water; therefore, when the acoustic energy enters the bottom at some angle, it refracts (i.e., curves) in the sediment. The ray paths begin to rapidly curve upward until they leave the bottom (the sediment) some distance away from the entry point. Depending upon the characteristics of the sediment, one of three things will occur: no loss in energy, some loss in energy, or substantial loss in energy. This energy loss will be dependent upon the characteristics of the sediment and the frequency of the acoustic energy. High frequencies generally suffer much greater loss than lower frequencies.

Ambient Noise

Noise is unwanted signal. If we go back to our example of a discussion between two people (you and George) in a room, the acoustic signal sent from George to you is signal. The signal is desired, and you are making efforts to hear and understand what George is saying (the signal). If we now fill the room with many other people, all of whom are talking to each other, then you can hear their voices, as well as George's. However, the acoustic signals sent by the other speakers in the room are interfering with the signal from George. Those interfering signals are, by definition, *noise*. You ability to hear what he is saying is reduced because of the noise everyone else is making. Exactly the same concept applies to the problem of underwater acoustics.

Our example now will relate to naval acoustics. The "speaker" is a submarine, which creates the signal (sound); the "listener" is the sonar system attempting to detect (hear) the submarine. All other acoustic signals in the ocean are considered noise. This noise, sensed by the detection device, is called *ambient noise*.

Ambient noise is primarily due to three different mechanisms: sea noise, biological noise, and man-made noise. Sea noise is caused by wind driving the surface of the sea into waves of all sizes, rain, hail, typhoons, and other weather-related phenomena. Biological noise other than that produced by whales is caused by marine life, including shrimp and other living organisms. Man-made noise is caused by all the machinery that man is operating in the oceans, such as ships, oil exploration equipment, oil platforms, and other machinery. The addition of these three components comprises the ambient noise.

Biological noise is generally of high frequencies; weather noise is also of higher frequencies, and man-made noise is generally of low frequencies. The principal component of ambient noise that affects the detection of submarines is man-made noise.

The principal source of man-made noise is from ships at sea. At any given moment approximately 16,000 ships are at sea, from a worldwide population of about 40,000 ships. These include cargo vessels, oil tankers, commercial fishing boats, and many others. The noise they make is generally due to engines and propeller movement.

There is also a special problem with all ships: they manufacture electricity using generators. Depending upon the country of origin these generators make alternating current at a specific frequency. All the Western Hemisphere, Japan, the Philippines, and other parts of Asia use 110 V and a frequency of 60 Hz. In the rest of the world, the frequency is 50 Hz.

All electrical equipment used throughout the world, whether on land or at sea, is extremely sensitive to the frequency. Variations in amplitude (voltage level) are quite acceptable. Voltage in the United States, for example, is 110 V at 60 Hz. The voltage level varies between 100 V and 120 V, and the electrical equipment is designed to allow for such variations. But the frequency is exactly 60 Hz. This is carefully controlled. Thus, generators on board United States ships operate at 60 Hz, and most the rest of the marine world operates at 50 Hz. The term “ships” also includes submarines, an important piece of information.

Additional engineering equipment on board makes noise as well. Large rotating machinery that is attached to the hull vibrates, and that vibration is transformed into acoustic energy that is transmitted into the sea. The noise output of a surface ship is dependent upon the type of equipment it has and how the equipment is mounted within the ship. Due to the operating equipment of the ship, the level of the generated acoustic signal will be speed dependent. That is, the level will increase as the ship goes faster or encounters heavy weather.

The same physics of noise generation on surface ships is also true for submarines. U.S. submarines take considerable care in attempting to quiet the operating equipment on board. Soviet submarines operate equipment that generates 50 Hz, and the same equipment generates signals that are transmitted into the ocean at 50 Hz. Because of the characteristics of rotating machinery, harmonics are also generated at 100 Hz and 150 Hz. The other equipment operating on board ships, including submarines, generates noise at many different frequencies.

Submarines operate at different depths and go where they choose for different operational requirements. Surface ships generally follow relatively well defined paths, commonly called *shipping lanes*. Ships are never uniformly distributed over the surfaces of the world's oceans. In fact, the ships are usually distributed in well-known shipping lanes.

Both the economy and the weather dictate the paths for such shipping lanes. The shipping lanes in the Northern Hemisphere generally go slightly farther toward the north in the summer, closely approximating the Great Circle Routes between major ports. In the winter, the same lanes shift slightly southward. Although this shift in a southerly direction increases the distance between ports, it also decreases the heavy seas that the ships encounter in the winter storms. This maneuver lessens wear and tear on the ships and allows them to proceed a little faster.

The signals from the surface ships propagate throughout the ocean, generally in the band from 25 Hz to 600 Hz. These acoustic signals are heard by the receiving sonar; since the signals are unwanted, they are considered to be noise. The received noise coming from the ships is unequally distributed over the ocean surface, and the noise from each individual ship propagates to the receiver and undergoes different transmission loss. Therefore, we can see that the *ambient noise at the receiver is directional* . This one fact alone allowed the design of equipment that permitted detection of submerged submarines at amazing distances (hundreds, even thousands of miles) from the receiver.

Receivers

In human beings, the ear is the organ that receives acoustic information. Equipment has been manufactured that is sensitive to acoustic stimuli; this equipment is called the *transducer*.

Transducers have different characteristics. They have a basic sensitivity that determines the level of signal that they can detect. Transducers are also frequency dependent. Building transducers that are sensitive to low frequencies is difficult; it is much easier to build transducers that are sensitive to high frequencies. However, in ASW, we are interested in relatively low frequencies (remember the example of the whales). Thus, we immediately encounter some practical engineering problems about which equipment to use in attempting to detect submarines.[2]

A single transducer may be relatively small—about the size of a baseball or an apple. Its small size means that the transducer is extremely sensitive to very high frequencies, which means that the wavelength of the signal is very small in comparison to the characteristic size of the transducer. The frequencies of interest in ASW have wavelengths that are much larger than the characteristic size of the transducer, which means that the transducer is essentially an omnidirectional device. It is sensitive to all signals, regardless of the direction from which they arrive. This high sensitivity is discouraging, since it would be valuable to have a sensor (the transducer) be able to sense the angle of arrival. This angle of arrival could be measured either from the vertical or in azimuth. This task is impossible with a single sensor because of basic fundamental physics; however, a branch of mathematics called *array theory* will allow us to overcome this problem.

Array theory is complex and uses sophisticated mathematics. I will not discuss details here. There are, however, some things that array theory can do that are critical to the understanding of sonar receiver development. Array theory is based upon the intriguing fact that it is possible to establish a geometrical pattern of omnidirectional receivers that will allow the angle of arrival to the geometrical array to be ascertained. One of the geometrical patterns that has this property is a straight line. If you place a number of omnidirectional receivers (say, 10) in a straight line and properly connect them together, then it is possible to determine the arrival direction of the incoming signal with substantial accuracy. In the ASW world, omnidirectional receivers are called *hydrophones* (sensors that can detect acoustic vibration in the water. When omnidirectional sensors are placed in a geometrical pattern, the pattern is called an array. An array where all the sensors are in a straight line is called a line array. In ASW the line of hydrophones is called a *towed array* because the line array is towed in a straight line behind a ship that is moving slowly forward (approximately 3 knots, just fast enough to make the array become relatively straight and also be able to maintain control of the ship).

Now, a little more complexity will allow us to understand some of the remarkable characteristics of towed arrays, or, as they are called in the

---

[2] It is not our purpose to discuss the engineering problems associated with building receiving sonars, but we do need to discuss the issues of directional sonars.

Navy, *tails*. All the hydrophones are wired together, and the signals they generate are brought back on board the ship through the cable pulling them. On board ship, the signals are subjected to several different mathe-matical processes (signal processing) that are done by using sophisticated computers. Because of the speed and capability of these computers, several things

> **Tails and Tales**
>
> Within this book are tales about tails. Since their initial development by Marvin Lasky of the Office of Naval Research in the late 1950s, towed arrays have been called "tails." *Towed arrays are a number of hydrophones placed in a line with varying distance intervals between them;hydrophones are microphones that can operate underwater.*
>
> The specialized electronic equipment of that time (and which eventually evolved into computers) were called *beamformers*. They allowed literally hundreds of beams to be formed simultaneously. Each beam could then be observed and analyzed. Although the hydrophones as fundamental acoustic detection sensors are omnidirectional in character, the linear placement of these hydrophones with variable distances between them allows the "beams" to form electronically. These distances are carefully calculated to allow the towed array to have certain desired operating characteristics.
>
> Towed arrays are pulled behind submarines and surface ships. The first towed arrays were only several hundred feet long; they steadily grew in length to arrays of over three *miles* long. Their depth in the ocean could be controlled in several ways, including the amount of tow cable that was unspooled from the winch and the speed at which the ship moved. The development of towed arrays and their use as detection sensors in the world of antisubmarine warfare were an integral part of the history of the LRAPP/ASW Environmental Acoustic Support (AEAS) programs.

can be done with the signals.

Using the computer, the signals may be treated so that it is possible to look in any desired direction, all the time. The array pattern, which is a direct result of the mathematics of Array Theory, consists of beams, nulls, and side lobes. The number of beams is directly related to the number of array elements, i.e., the hydrophones, so 10 hydrophones will allow the formation of 9 beams. Further, thanks to array theory, the relative amplitudes of the hydrophones can be adjusted so that the beam of interest, i.e., the one looking in the direction of interest, can have a magnitude that is arbitrarily larger than all the beams looking in other directions. This combination of techniques has the practical result that signals arriving from directions other than the main beam direction will be substantially attenuated.

Further still, there are nulls, which are very low levels of arriving signal attenuation. Additional advances in extensions to array theory allow the nulls to be forced to look in any stipulated direction. That means that if a high level source of noise occurs in one particular direction, you can “null” it out.

Unfortunately, all these functions work wonderfully well in theory: all the arrays are straight; all the hydrophones work the same way; nothing goes wrong. In practice, it doesn’t work that way. As the towed array is pulled through the water, it follows a sinuous motion. The hydrophones are not identical; in fact, once in a while, some of them don’t work at all. The water moving along the surface of the cylindrical array makes noise that interferes with the signals being received. These are real-world engineering problems. However, if you know the shape of the tail at any instant of time, many of these errors can be mathematic-ally removed. The expansion of computer capability into small and inexpensive devices has allowed most of these problems to be solved to at least the level required for satisfactory engineering results.

Summary

Physics and mathematics have been addressed with regard to the issues related to detecting submarines. The ocean environment, including the speed of sound, the bottom, and the ocean surface, has characteristics that have slowly been discovered and employed to allow the Navy to predict the performance of sonar systems. These characteristics are represented in mathematical formulas and programmed into computers.

The growth of computer capabilities, the knowledge of the ocean environment, and the application of array theory to towed arrays or tails, allow the dependable prediction of submarines detection at great distances.

All of LRAPP’s efforts were focused on oceanographic measurements, generation and validation of mathematical models, and applications to real ASW problems in support of the national policy of defeating the Soviet submarine threat through the world’s oceans.

FINIS